ROUTLEDGE LIBRARY EDITIONS:
COLD WAR SECURITY STUDIES

Volume 33

NONOFFENSIVE DEFENSE

NONOFFENSIVE DEFENSE

A Global Perspective

UNIDIR • UNITED NATIONS INSTITUTE FOR
DISARMAMENT RESEARCH

Routledge
Taylor & Francis Group

LONDON AND NEW YORK

First published in 1990 by Taylor & Francis Ltd

This edition first published in 2021
by Routledge
2 Park Square, Milton Park, Abingdon, Oxon OX14 4RN

and by Routledge
605 Third Avenue, New York, NY 10017

Routledge is an imprint of the Taylor & Francis Group, an informa business

British Library Cataloguing in Publication Data
A catalogue record for this book is available from the British Library

ISBN: 978-0-367-56630-2 (Set)
ISBN: 978-1-00-312438-2 (Set) (ebk)
ISBN: 978-0-367-62821-5 (Volume 33) (hbk)
ISBN: 978-1-00-311096-5 (Volume 33) (ebk)

Publisher's Note
The publisher has gone to great lengths to ensure the quality of this reprint but points out that some imperfections in the original copies may be apparent.

Disclaimer
The publisher has made every effort to trace copyright holders and would welcome correspondence from those they have been unable to trace.

Nonoffensive Defense

A GLOBAL PERSPECTIVE

UNIDIR
United Nations Institute for Disarmament Research

Taylor & Francis
New York • Bristol, PA • Washington, D.C. • London

USA	Publishing Office:	Taylor & Francis New York Inc.
		79 Madison Ave., New York, NY 10016-7892
	Sales Office	Taylor & Francis Inc.
		1900 Frost Road, Bristol, PA 19007-1598
UK		Taylor & Francis Ltd.
		4 John St., London WC1N 2ET

Nonoffensive Defense: A Global Perspective

Copyright © 1990 United Nations Institute for Disarmament Research

First published 1990
Printed in the United States of America

Library of Congress Cataloging in Publication Data

Nonoffensive defense: a global perspective / UNIDIR.
      p. cm.
  Papers presented at a symposium organized by UNIDIR! held Sept.
1989, in New York City.
  ISBN 0-8448-1683-3 (HC)
    1. Deterrence (Strategy)—Congresses. 2. Offensive (Military
Strategy)—Congresses. 3. Armaments—Congresses. 4. Military
policy—Congresses. 5. World politics—1985–1995—Congresses.
I. United Nations Institute for Disarmament Research.
U162.6.N66   1990
355.02'17—dc20   90-11112
                                CIP

UNIDIR

United Nations Institute for Disarmament Research
Institut des Nations Unies pour la Recherche sur le Desarmement

UNIDIR is an autonomous institution within the framework of the United Nations. It was established in 1980 by the General Assembly for the purpose of undertaking independent research on disarmament and related problems, particularly international security issues.

The work of the Institute aims at:

1. Providing the international community with more diversified and complete data on problems relating to international security, the armaments race, and disarmament in all fields, particularly in the nuclear field, so as to facilitate progress, through negotiations, toward greater security for all States and toward the economic and social development of all peoples;
2. Promoting informed participation by all States in disarmament efforts;
3. Assisting ongoing negotiations in disarmament and continuing efforts to ensure greater international security at a progressively lower level of armaments; particularly nuclear armaments, by means of objective and factual studies and analyses;
4. Carrying out more in-depth, forward looking, and long-term research on disarmament, so as to provide a general insight into the problems involved, and stimulating new initiatives for new negotiations.

The contents of UNIDIR publications are the responsibility of the authors and not of UNIDIR. Although UNIDIR takes no position on the view and conclusions expressed by the authors of its research reports, it does assume responsibility for determining whether they merit publication.

UNIDIR

Palais des Nations
CH-1211 Geneva 10
Tel. (022) 734 60 11

Contents

Part V Defense, Disarmament, and the International Community

Preface

The conduct of forward-looking and long-term research to stimulate new initiatives for new negotiations is an essential part of the mandate of the UN Institute for Disarmament Research (UNIDIR). There has for some time been an increasing realization that the arms race, and especially the nuclear arms race, cannot be sustained by any political or economic rationale. The dramatic improvement in the relationship between the United States and the USSR and the concrete agreements this has yielded have had a ripple effect globally. The United Nations is engaged in the tasks of peacekeeping and peacemaking in regional conflicts. The concept of common security has made rapid advances. In the realm of ideas, a number of new concepts have gained currency in strategic thinking.

In the vigorous discussion on East-West issues, a group of proposals variously referred to as "nonoffensive defense," "nonprovocative defense," "alternate defense," and others has emerged. These proposals have been made with a view to achieving conventional stability. The maximization of the defense orientation of forces, it has been claimed, will reduce the danger of surprise attack, achieve security at lower levels of armaments, enhance confidence, and reduce suspicion leading to peace and stability. The structural incapacity to attack will eliminate threat perceptions. These claims have been supported by some, questioned by some, and refuted by others. Some have found in the concept a partial answer, and they point to its neglect of the question of nuclear deterrence. Is the distinction between offense and defense a matter of military strategy or a political attitude? The debate goes on, and the definition of nonprovocative defense is coming into sharper focus through the valuable work of many writers. The concepts have nonetheless had an indisputable impact, and their inclusion in the policy statements of some world leaders is evidence of this.

The proposals I have referred to emerged in the context of the East-West rivalry in Europe. Although their feasibility as disarmament proposals and confidence-building measures deserves continued analysis, it would also be interesting to transpose these ideas into other regional contexts. In a highly integrated global system, the stability achieved in one region will always be fragile as long as instability persists elsewhere. The task of finding solutions to regional conflicts and of achieving stable security regimes in different parts of the world has acquired greater urgency with the improvement of the superpower relationship. Nonoffensive defense is one possible avenue. We would be failing in our duty as a research Institute if we did not at least explore that avenue. This, then, was the background of the subject for a symposium organized by UNIDIR in September 1989 in New York City.

UNIDIR was fortunate to find in the United Nations Association of the United States (UNA USA) an enthusiastic partner in the organization of this symposium. I thank Edward Luck and Jeff Laurenti for their remarkable organizational skills in ensuring that our transatlantic cooperation achieved a successful outcome. I also thank Dr. Albrecht von Muller of the Max Planck Society, one of the foremost writers in the field of nonprovocative defense, who joined UNIDIR and UNA USA as a cosponsor. Finally, I thank the Volkswagen Foundation of the Federal Republic of Germany whose grant contributed toward holding this symposium.

The publication of the papers presented at the symposium as well as some papers reflecting on the symposium itself in retrospect provides a useful attempt to broaden the discussion. This book is divided into six parts: Part I attempts, through different approaches, to define defensive strategies and examine their credibility; Part II examines these strategies in the European bipolar context; Part III looks at the situation in Northeast and Southeast Asia and the possible applicability of defensive strategies; Part IV examines conflict situations in the Middle East and Central America; and Part V takes a global view of the problem. The concluding part reflects on the discussions that took place at the symposium, indicating possible future directions for research.

The views expressed in the chapters are the responsibility of the authors and are not necessarily shared by UNIDIR. However, UNIDIR does commend this volume to the attention of its readers.

Jayantha Dhanapala,
Director

Part I

Paradigms of Defensive Strategies: Concept and Credibility

Chapter 1

Conventional Stability and Defense Dominance

Albrecht von Muller

Over the last few years, conventional stability has become a fashionable issue. But this new prominence is not yet matched by conceptual clarity. Many still confound the notions of equilibrium or balance, that is, forms of symmetry, with the notion of stability. But in order to achieve the latter, much more is required than just balanced military potentials.

The best example of this is shown by two cowboys slowly approaching each other in the dusty mainroad of Dodge City for the final shootout. The situation is characterized by perfect symmetry (equal weaponry, equal ammunition stockpiles, equal capabilities, etc.). Yet, the situation is highly unstable. The reasons are the high bonuses for pre-emption and attack.

In order to achieve a regime of conventional stability, we must get to the inverse situation, that is, the bonuses for pre-emption and attack must no longer be decisive, and the defensive capabilities of both sides must clearly exceed the offensive capabilities of the respective opponent. Such a regime can be defined as a "mutual defense dominance." It requires an in-built mechanism that shifts the force ratio in favor of the defending side, no matter which side tries to attack.

It is the purpose of this chapter to provide the conceptual framework for discussing conventional stability issues and to give an idea of how a regime of defense dominance, applied to Central Europe, might look.

A METAPHOR: THE SWORD FIGHTERS AND THE HORNETS

Today's conventional forces in Europe can be compared to ancient fighters equipped with swords and shields. There exist offensive and defensive capabilities, and both are genuinely intertwined. Due to the fact that initiative and surprise effects play a important role, even a perfect symmetry of those forces would not provide stability.

For conventional stability the military potential of both sides must contain components that are more efficient in defensive than offensive mission. In our example such a factor might be a swarm of hornets that defend their own territory and attack any intruder immediately, but do not cross the border into foreign terrain. This would mean that the attacking sword fighter would be confronted not only with his own mirror image but, in addition, with a swarm of highly aggressive hornets; that is, there would exist a factor that shifts the balance in favor of the defending side.

Beyond these basic stability features, our examples also allows us to demonstrate the crucial military mechanism of a regime of defense dominance. Not only does an additional factor come into play, but the new combination also constitutes a fundamental dilemma for the attacking side. If the attacker tries to protect optimally against the hornets, he becomes a poor sword fighter. If, however, he concentrates on the fight with the opposing sword-fighter, he is easy prey for the aggressive insects.

As these effects apply in both directions, there exists a regime in which neither side has any problems in the role of the defender but would maneuver itself into a position of devastating inferiority if it would ever try to attack. And that is exactly what conventional stability is all about: war prevention through devastating prospects for the side that would start the war.

DEFINITION OF CONVENTIONAL STABILITY

The notion of stability stems from physics. There we call a dynamic system stable if disturbances activate correcting forces, which bring the system back into its original state. The best example is the marble in the bowl which—as long as it is not thrown out altogether—will always roll back to the lowest point. Transferring this notion of dynamic stability to the sphere of security policy, we get a clear picture of what conventional stability is about: The goal is a military setting in which disturbing impacts—such as crises, accidents, or hostile intentions—do not lead to escalatory dynamics but are contained and regulated back so that peace can be maintained.

In general, there are two ways in which war can break out. One is intentional aggression and attack as committed by Hitler in 1939. The second is the process of crisis escalation in which military necessities take over and dominate the political decision-making process. Many historians cite the last weeks before the outbreak of World War I as an example for such a process. A conventional stability regime has to make both ways of war as unlikely as possible. It is very important, however, to understand that there can arise conflicts of interests between these two tasks. If one concentrates only on deterring intentional aggression, one easily takes measures that increase the probability of crisis escalation. In contrast, exclusive emphasis of the avoidance of escalatory dynamics may weaken deterrence against intentional aggression.

The virtue of a comprehensively defined notion of stability is that it forces us to consider both dangers and thus leads to a comprehensive security and defense policy. Three formal definitions of conventional stability can be offered; verbal, mathematical, and operational.

The verbal definition reads as follows:

> Conventional stability exists only if the robust defense capabilities of both sides clearly exceed the offensive capabilities of the opponent. Only if this is the case—and clear to both sides—can the conventional defense preparations fulfill their function, i.e., prevent the outbreak of war in case of crisis, accidents, or hostile intentions.

The mathematical definition reads as follows:

$$D_x - P_y >> O_y$$

(D = overall defense capability; 0 = overall offense capability; P = preemption bonus; x, y refers to the contrahents)

The operational definition reads as follows:

> The Force Ratio Development Functions (FDFs) intersect, i.e., the initial advantages that the attacking side can gain by surprise effects and by the capability to decide when and where to attack, are more than compensated in the following phases of the conflict.

This operational definition is the most informative one of the three and should, therefore, be explained in some detail. A FDF describes how the overall force ratio changes during conflict, that is, it describes the path along which the force ratio changes from the initial ratio at the outbreak of armed conflict to the final ratio where one side is worn down completely.

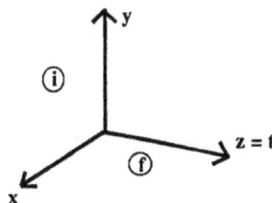

(x = aggregated strength of A; y = aggregated strength of B; z = time; i = initial force ratio; f = final force ratio; an intersection of the FDF with the plane (x = 0) means that A has lost, an intersection with the plane (y = 0) means that B has lost.)

In order to simplify things, we reduce our three-dimensional phase space to a two dimensional one by dropping the time axis. (Figuratively speaking, we turn the coordinate system in such a position that the time axis or runs exactly to or exactly away from the observer, i.e., it disappears.) Now we introduce two different FDFs, one representing the case that A attacks, the other that B attacks. By doing this we can identify the three basic stability variants:

SV 1: Full superiority
      (one side wins, no matter whether attacked or attacking).
SV 2: Instability
      (both win as attacker, lose as defender).
SV 3: Stability
      (both win as defender, lose as attacker).

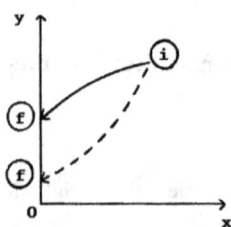 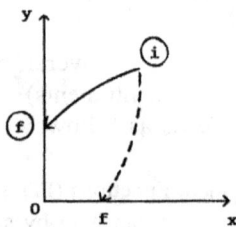 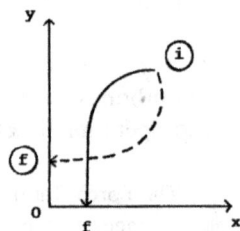

SV1   SV2   SV3

From these extremely simplified models, we can see some interesting insights in the optimization of conventional stability.

Unwise reductions can lead to instability:

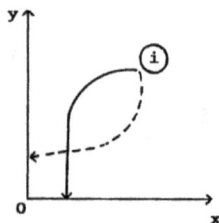 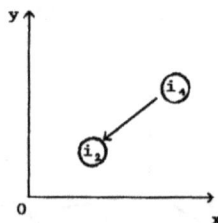 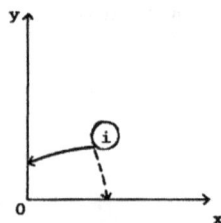

Initial Situation   Simplistic Reductions   Deteriorated Stability

In order to make reductions useful, they must be combined with a changed FDF topology, that is, the bonuses for pre-emption (P) must be minimized and the exploitation of defense advantages must be optimized (flattening of FDF in favor of defender):

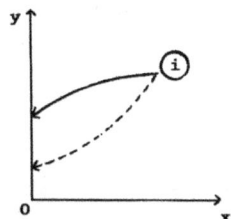

| Initial Situation | Modified Force Struct. | Stab. at Reduced Force Levels |

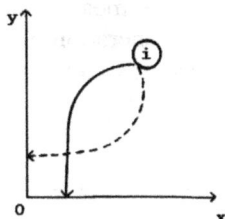

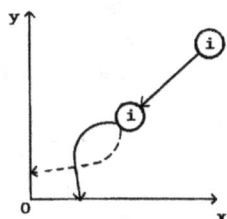

CRITERIA AND PURPOSES OF CONVENTIONAL STABILITY

As we have seen, conventional stability is a fairly complex phenomenon. It is, therefore, useful to disaggregate it in its basic components. This can be done best along the time axes, and by doing this we can identify the four following substabilities:

1. Pre-emption stability: The possible bonuses for pre-emption must be minimized. This can be done by three means: (a) the reduction of the number and relevance of targets for pre-emption, (b) improved protection for the remaining, inevitable targets, (c) the limitation of weapon systems that can be used for pre-emptive missions.
2. Battlefield stability: The exploitation of the advantages of the aggressor on the battlefield must be inhibited. This means that especially surprise effects and local superiority must be denied. This can be done by limiting offense prone mobility, massing, and momentum. It also can be achieved by geographical and/or numerical limitations of amunition stockpiles, mobile bridging equipment, autonomous logistics, etc. At the same time the structural advantages of the defender must be systematically exploited. This means that there should be no limitations to attrition-oriented weapon systems and units that do not possess strategic mobility. This includes especially modern mine technology, barriers, terrain preparation, and antitank and antiair PGMs as long as they are not mounted on strategically mobile platforms.
3. Force generation stability: The potentials for force generations must be controlled and shaped in a way that no significant superiority can be achieved at any time or by any mobilization pattern.

4. Arms race stability: The overall military setting must make sure that the acquisition of additional offensive capabilities is by far more expensive than the means to compensate them.

These four criteria relate to four different time horizons. The pre-emption calculus uses minutes and hours as measurement units, the battlefield calculus days and weeks, the force generation calculus weeks and months and the arms race calculus finally months and years.

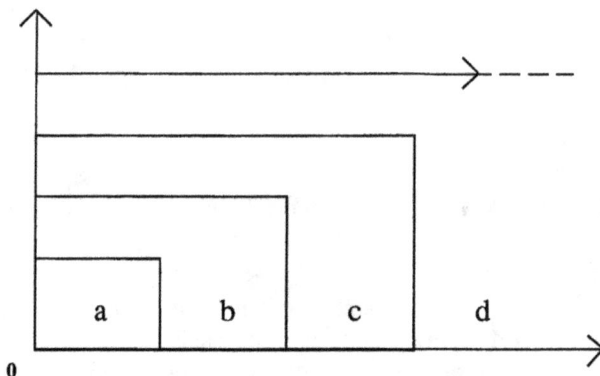

A regime of conventional stability in the fullest sense exists only if the overall goal of defense dominance exists regarding all four of these time horizons. This statement also makes clear that it is not possible to achieve conventional stability forever, but that its maintenance is a continuous political task.

As to the purposes of conventional stability, we can identify three basic functions:

1. To avoid unintentional war, which might occur if, in case of a political crisis, military necessities take over and dominate the political decision-making process.
2. To avoid intentional war, that is, to signal to the opponent that his own offensive capabilities are by no means sufficient for successful strategic attack and invasion.
3. To pave the way for a reduction of the conventional force levels (which is impossible in an ambivalent or defense-dominant regime) and to allow a shrinking of the nuclear component to its politically required minimum.

These three purposes and the four stability criteria noted above relate to each other in a specific way:

Criteria		Goals	
Pre-Emption Stability		Avoidance of Unintentional War (Crisis Escalation)	
Battlefield Stability			
Force Generation Stability		Avoidance of Intentional War (Planned Aggression)	
Arms Race Stabilty		Reduced Defense Expenditures	

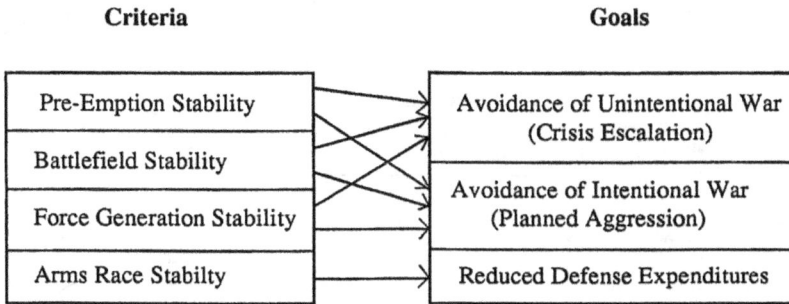

THE STATUS QUO IN EUROPE AND HOW WE GOT THERE

Analyzing the present military confrontation in Europe from this stability-oriented point of view, the picture is quite bizarre. Both sides spend much money for their conventional forces (an order of magnitude more than for the nuclear forces) and the spending has been roughly balanced for at least two decades. Yet the stability rests mainly on political factors and on the existence of nuclear weapons as an ultimate threat. If we just imagine a very hostile, aggression-oriented political climate, combined with the absence of nuclear weapons, we see immediately how extremely little the huge and expensive conventional forces of both sides contribute at present to our net stability.

So, from a stability point of view, one could say that a most bizarre misspending took place over the last decades on both sides. The reasons for this are many, the most important probably being the mixing up of the notions of symmetry and stability. Another important factor is probably the thought patterns and modeling tools inherited from World War II.

The most advanced form of fighting during World War II was the Guderian type of highly mobile, breakthrough-oriented tank units. The modeling approach for that is the Lanchester Square Law, which reflects the basic features of a confrontation of those forces quite well. Until today it has been the core element of military force planning on both sides. Yet, this constitutes a problem. Using this modeling approach, there exists an built-in bias for highly mobile, heavy armored units, as both sides have them today. And due to the intrinsic force concentration and breakthrough capabilities of these forces, even a perfect equilibrium does not provide stability (see Fig. 1.1). Therefore it was possible that both sides spent enormous resources over decades for those forces without any progress regarding the achievement of conventional stability in Europe.

One has to understand this self-immunizing feedback loop of thought pat-

Practice   Theory/Modeling

```
                        ┌─────────────────────────────────┐
                        │        Emphasis on              │
                        │ high mobility and great fire power │
                        │ leading to very expensive platforms │
                        │      that require heavy armor    │
                        └─────────────────────────────────┘
```

Emphasis on
high mobility and great fire power
leading to very expensive platforms
that require heavy armor

Units with a
high degree of functional complexity
leading to a nonlinear efficiency decay
beyond a certain threshold of losses

Crucial role
local superiority and rapid breakthrough
as correctly reflected in the
Lanchester Square Law

Strong incentives
to concentrate forces and to
push the enemy beyond this
threshold in a rapid blow

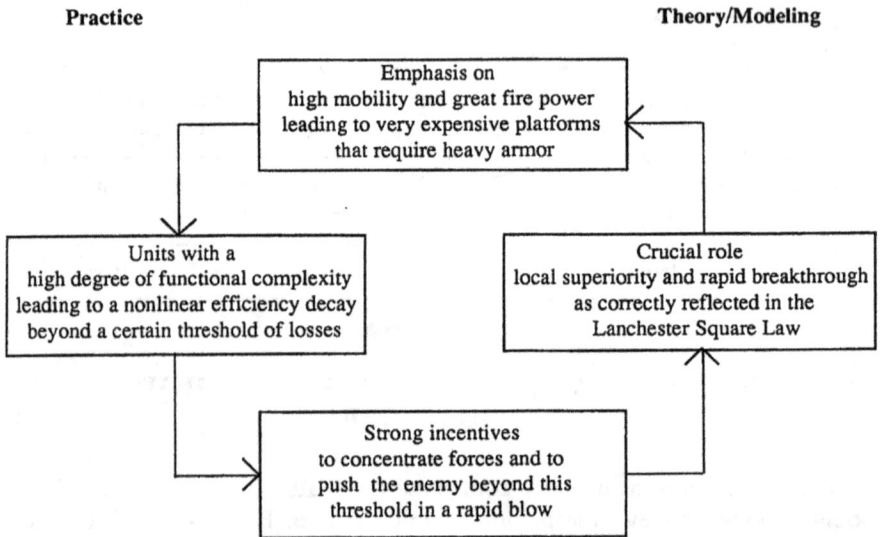

Figure 1.1 The logic behind today's land forces and their inherent instabilities.

terns and force developments on both sides to understand how we got to the status quo and how we could overcome it. I became aware of this fatal self-immunization only when we started to propose modified force structures, drawing on the synergism between the traditional, heavy armored, highly mobile units and new, specifically attrition-oriented components. When these proposals were analyzed by various MODs, the results were always very meager and by analyzing the analysis we found that the modeling tools simply did not allow a correct representation of the new, attrition-oriented components and their synergism with the traditional units.

This fact also explains the opening positions for the Vienna talks. Unless one really digs into the problems of conventional stability and develops new modeling tools, one can just demand an equilibrium of today's forces, taking into account the required force-to-space ratios. And that is exactly what NATO and WTO did and what brought us to the not very imaginative opening position for the Vienna talks. It will be a step forward if along these lines a regime of equal ceilings can be achieved. But nobody should misunderstand it as a regime of conventional stability, which it will not yet be.

THE BASIC FEATURES OF A CONVENTIONAL STABILITY REGIME

How could this self-perpetuation of unstable force structures be broken up; what would a sword-fighter-plus-hornets regime look like? When trying to answer these questions, we must focus on the central front, not because the flanks or other regions are regarded as irrelevant, but in order to elaborate the basic features of a stability regime. With certain adaptations the same logic can then be applied to the northern and the southern flanks as well as to other regions.

Obviously, the heavy armored, highly mobile units would be the sword-fighters of our example. But who could fulfill the function of the hornet? The answer is modern barrier and infantry units that possess only local mobility and are equipped with new types of passive ammunition and "close interdiction" capabilities that allow the engagement of force concentrations over ranges of 5 to 20 kilometers with a high degree of efficiency. What we are talking about are attrition-oriented forces without strategic mobility but with high efficiency against both dispersed and massed forces. The basic idea of the proposed conventional stability regime is, therefore, to overcome the present monoculture of heavy armored, highly mobile forces that exists on both sides.

Such mixed force structures would improve conventional stability drastically. The defending side could utilize all its forces while the attacker could bring only his mobile assets into play. This means that the general staff of both sides would have to tell their political decision makers: "If we stay home, there is no problem whatsoever. If, instead, we take over the role of the attacking side, we would maneuver ourselves into a position of devastating inferiority." And that is the core of the conventional stability problem: no chance whatsoever for a successful aggression combined with mandatory military incentives to stay home and wait even in the case of acute crisis.

The next question is how one could make sure that the air forces cannot topple this stable regime for the land forces. At present, the air forces provide by far the largest bonus for pre-emption and are thus the most destabilizing factor in case of crisis. In any regime of conventional stability, the air forces have to have, therefore, a fairly different shape and function.

Today NATO needs strong air capabilities to compensate for a huge quantitative superiority of land forces by the WTO. But the maintenance of these air capabilities has an enormous price for NATO. We cannot deny identical air capabilities to the opponent and this, in turn, constitutes a major disadvantage for NATO. Assuming that NATO will never be the attacking side, the very large bonus of pre-emption will be gained by the opponent.

The only way out of that dilemma in a future stability regime is (1) to deny quantitative superiorities through equal ceilings, (2) to reduce the attack-capable land forces to an extent that strategic echeloning is no longer possible, (3) to redefine the function of the air force, substituting the mission of deep penetrations for the mission of denying air superiority over one's own territory, and (4) to reconfigure the air forces accordingly, paying specific attention to reduced pre-emption capabilities and improved protection (for example, by putting more emphasis on vertical takeoff planes, etc.).

This is a long list and difficult to fulfill, but nobody pretends that conventional stability could be achieved easily and without substantial changes of the present military setting in Europe. These arguments bring us to the architecture of a conventional arms control regime. But before we dig into that, we should come back to the question of force modernization, that is, how the attrition units advocated above could be equipped and composed.

OUTLINES OF A STABILITY-ORIENTED FORCE MODERNIZATION

The goal is to complement today's forces with new, attrition-oriented units that are highly efficient in the delay and defense battle but of little use for strategic offensive missions and invasions. One can argue that basic technological trends at present favor such a strengthening of defense (see Fig. 1.2). But there is nothing automatic about this development, that is, it is the responsibility of the political decision makers to select those options that lead to a nonambiguous enhancement of defensive capabilities.

The best example for such a stability-oriented modernization is probably a new weapon system that comprises two components: target acquisition by a network of dug-in seismic sensors, and stochastically mobile mines. This system configuration is based on two ideas. One is to utilize seismic signals for a very precise, reliable, real-time target acquisition right from the border. The sensors are dug in at a depth of three to four meters, that is, artillery resistant, and the distance between each of them is 200–500 meters. A grid of these very cheap, redundantly netted sensors could be placed in all tank-apt strips along and close to the respective border.

The second idea relates to the ammunition. The mainstream thinking today goes in the direction of intelligent submunition, a trend that raised the price per shot drastically. Here, instead, it is proposed to serve the targets with stochastically mobile mines. These are sort of iron footballs, filled with roughly 15 kilograms of explosive, that bounce around on the battlefield like a fire cracker. Propelled by small rocket motors, they run about eight or ten lanes, 300–400 meters each. As soon as they hit a big piece of metal, they are detonated by a magnetic charger. By their movement they increase their kill probability, compared to static mines, by orders of magnitude. For the delivery of these mobile mines, cheap, nonintelligent artillery rockets with ranges of 5–60 kilometers are used.

The combination of these components has various virtues, military-technical, spending wise, and political. The military virtues are basically four:

1. It optimally complements dispersed infantry and barrier units creating a insoluble technical dilemma for the attacker. The more he disperses,

Four Basic Analytical Categories of Military Action

1. "Fire" (as explosive and its allocation)
2. "Movement" (on land, in the air, and in water)
3. "Protection" (through armor, deception, ECM, etc.)
4. "Reconnaisance" (early warning and target acquisition)

Technological Progress in these Four Fields since World War II

1. approx. factor 10–100
2. approx. factor 3–7
3. approx. factor 3–10
4. approx. factor 10–1,000

Asymmetrical Tasks of the Attacker and Defender

	Attacker's Role	**Defender's Role**
Advantage	Momentum of surprise and local superiority	Combat on familiar and prepared terrain
Disadvantage	Forward transport of equipment and personnel	Need for time- and space-covering defense

Resulting Asymmetrical Relevance Profiles

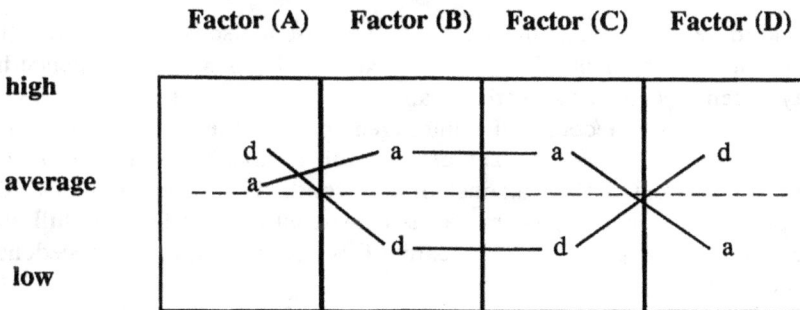

Normal distribution : _ _ _ _ ;   Attacker : a - a ;   Defender : d - d ;

Figure 1.2 Outlines of a stability-oriented modernization

the more he comes in contact with infantry and barrier units. The more he tries to mass, the more he becomes an optimal target for these stochastic mines. Against an enforced motorized battalion, which has a density of roughly 50 vehicles per square kilometer in the assault area, one would fire a salvo of 50–70 of these stochastically mobile mines at the same time. Given a kill probability of about 15 percent per mine, it is easy to calculate the enormous attrition rate that can be achieved by such a salvo. As the single mine has no target acquisition capability and is just moving stochastically, its price can be fairly low. Cost estimates go for prices between $2,000 and $3,000. This means that an extremely favorable cost to attrition ratio can be achieved for the defending side.

2. The second military virtue is the fact that the target acquisition is extremely robust. In order to create the seismic signal of a 40-ton vehicle on tracks, one has to have a 40-ton vehicle on tracks. Electronic mimicking and other tricks simply do not help, nor is it possible to disturb the target acquisition procedures by ECM or other means. Of all physical signals, the seismic signals is probably the most robust and unbetrayable one.

3. The third military virtue relates to the fact that such a system does not provide any targets for a pre-emptive strike. The only way to take out a sensor is a full artillery hit. But this is meaningless as the network is redundantly netted and operates almost at maximum efficiency even with only half of the sensors.

4. A positive side effect of these sensoring capabilities is that artillery positions even in a depth of up to 40 km can be immediately located, thus allowing for immediate counter action.

Beyond these military-technical virtues, the system also is a most interesting step in the development of intelligent weapons. Up to now intelligence has always been placed in the warhead and thus has been lost with every shot. Here, instead, we decouple the intelligent part of the system (the sensor network) and the explosive. Target acquisition capabilities, therefore, can be used again and again—against first, second, third echelon forces. The only part of the weapon system that is consumed is the relatively dull and, therefore, cheap explosives. This feature is the reason for the unprecedented cost efficiency.

The third category of virtues relates to the political signal. The proposed weapon system makes very clear that one would defend oneself directly behind the border with maximum efficiency. At the same time, it signals the opponent that one has no offensive intentions whatsoever. The system is unusable for any forms of attack as there is no target acquisition capability on foreign terrain.

Summing up, one can say that this is an example of a modernization option that makes an intelligent use of modern technology by translating it into a weapon system that exploits the specific advantage of the defending side (to be on one's own terrain and to be able to prepare it).

There are many other examples of modernization options that fulfill military-technical as well as stability criteria and fit the overall political goals. In general, we are talking here about an approach that could be called "close interdiction." The idea of "deep interdiction," that is, the idea to interdict second echelon forces already at a depth of several hundred kilometers, was militarily attractive but counterproductive regarding the stability features of the overall regime as these capabilities drastically increased the possible bonuses for pre-emption. At the same time their political signal was ambivalent, to say the least. If, instead, we focus on ranges between 5 and 60 kilometers, the picture changes quite drastically. Those weapon systems do not constitute a pre-emptive threat for the opponent nor do they offer targets for pre-emption.

A side effect is that they are militarily more efficient. If one tries to engage land forces already at a depth of several hundred kilometers, the opponent has a wide spectrum of options for countermeasures. He can mount additional shields on top of his vehicles; he can saturate the target area with shield targets like old trucks; and he can do so right in the realm of electronic mimicking. According to the "close interdiction" philosophy, instead, one goes for ranges where the battlefield constraints already apply. If the opponent has heavy shields mounted on top of his vehicles, he would be a lame duck in mobile warfare. The saturation of the target area with old trucks and other dummies also is no longer feasible so close to the battlefield.

So, in sum, we see here the outlines of a modernization philosophy that basically goes for a combination of three components:

1. Heavy armored, highly mobile units of the traditional kind for counterattack missions, but at significantly reduced force levels.
2. New light infantry and barrier units specialized for attrition warfare close to the border.
3. New close interdiction systems that constitute a functional complement to the infantry and barrier units, making sure that dispersing is punished as well as concentrating.

Altogether, we are approaching here a twofold synergism, one between a counterattack and attrition-oriented components and one inside the attrition component between its two basic elements (see Fig. 1.3).

It is not possible to describe here all details of a future stability-optimized conventional regime, but it should have become clear that:

```
┌─────────────────────────────────────────────────────────────┐
│            Impossibility to optimize on counterinfantry warfare │
│ by very heavy armor, extensive antibarrier equipment, specific tactics, etc. │
└─────────────────────────────────────────────────────────────┘
```

ATTRITION FORCES
(emphasis on barriers, infantry,
and close interdiction)

infantry and barrier units

punishment
of dispersing

punishment
of massing

close interdiction capabilities

COUNTERATTACK FORCES

today's heavy armored,
highly mobile forces
with great fire power and
highly mobile logistics

```
┌─────────────────────────────────────────────────────────────┐
│                Impossibility to optimize on mobile warfare,     │
│ plus (a) gained time, (b) previous attrition and (c) disruption/exhaustion │
└─────────────────────────────────────────────────────────────┘
```

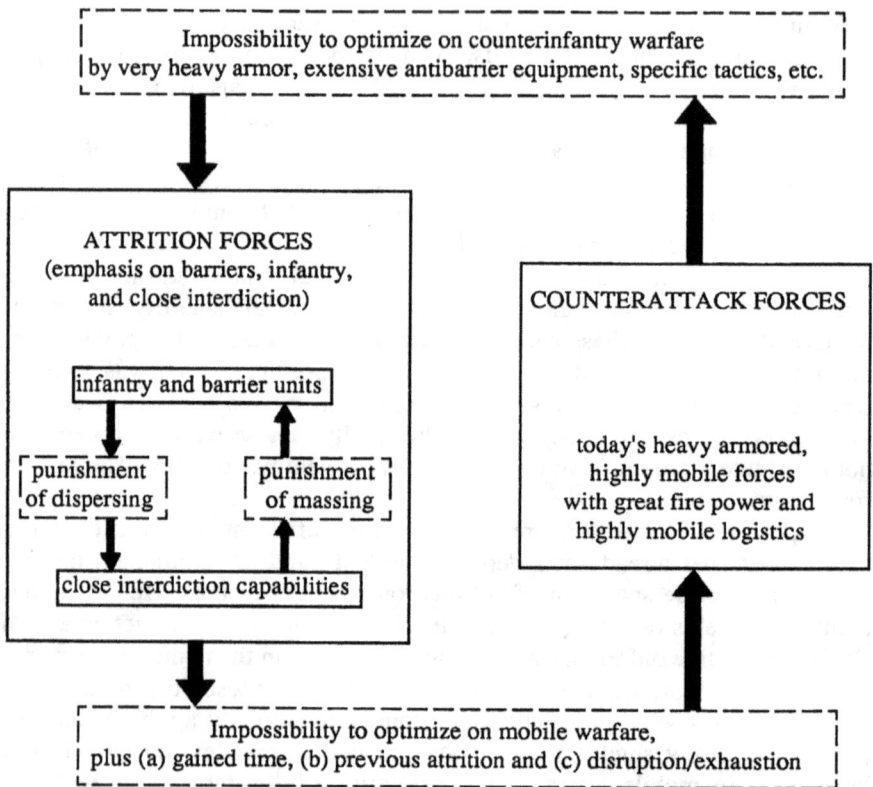

Figure 1.3 The two synergisms of an "Integrated Forward Defense"

We need a new approach to force modernization, one that is tuned to the
overall stability goals.

We need to integrate force modernization and arms control policies, and
in order to do this we need new evaluation criteria and modeling tools.

As this holds true for both sides, one should even consider elaborating these
criteria and assessment tools, at least in part, jointly, as the new goal is no
longer offensive capability and threatening force postures but the organization of significantly increased stability at substantially reduced force levels.

OUTLINES OF AN ARMS CONTROL REGIME

After defining and elaborating what conventional stability is and after discussing how modern technology might be utilized to approve the stability

features of the conventional confrontation, we must ask the question of what a related arms control policy could look like. It is important to make clear about which time horizon one is talking. We are talking about a regime for the year 2000, that is, the final outcome of the arms control endeavors in Vienna. Nevertheless the year 2000 is only a decade away and as we have long lead times, especially in arms procurement, we must conceptualize this goal now and start to work seriously for it if we want to achieve it roughly by the year 2000.

There are three core elements for a regime of conventional stability: (1) one relating to the land forces, (2) one relating to the air forces, and (3) one being the link to nuclear deterrence.

1. Concerning the land forces, it is important to reduce drastically the most offense-capable platforms. As we saw, the attacker has the fundamental task to invade and conquer foreign territory, that is, to move his personnel and weapons into foreign hostile territory. For this purpose he needs specific platforms, and the best avenue to cut down offense and invasion capabilities is to cut down these offense capable platforms.

   Nevertheless it is not advisable to abolish them altogether. They are needed not only for strategic offenses and invasions but also to restore territorial integrity, and the capability to do this will be a crucial feature of any future stability regime.

   Independent of this, it is important to maintain a complex threat for the attacker to avoid a counteroptimization. In the early stages of the debate on modified conventional force structures, some theoreticians have been arguing for almost exclusive reliance on PGMs. This approach would have led to major disasters already a couple of years later. As especially the appearance of active armor proved, it would have been a fairly easy task for the attacker to counteroptimize against such a technological monoculture. The question is not whether but how much heavy mobile units are needed for a efficient forward-defense posture.

   On the other end of the political spectrum, there have been and still are strategists that argue that NATO could not go significantly beyond today's force levels without getting below the "operative minimum," that is, without losing the capability for a coherent forward defense. This argument is correct if one assumes that there exist today's force structures. But the whole point is to modify the force structures in a way that specifically attrition-oriented capabilities are significantly increased while one can build down the ambivalent heavy armored, highly mobile forces. Therefore, the argument regarding the "force to space ratio" is correct, but it misses the point if drastic cuts on these mobile components are just interpreted as proportional reductions of the force to space ratio, and the complementary increase of attrition-oriented forces is not taken into account adequately.

The question is how much is sufficient for these counterattack capabilities. According to our analysis, roughly 50 percent of present NATO force levels regarding these offense-capable components are sufficient—given that these ceilings apply for both sides and that a complementary increase of the attrition-oriented components is realized. In concrete figures this would mean that both sides maintain roughly 10,000 main battle tanks and that the other major offense-capable platforms (IFV/APC, Arty, etc.) are reduced accordingly.

In addition to the overall ceilings, there must be density limits that make sure that neither side is allowed to make attack-capable force concentrations. Such density would be clearly preferable to a zonal approach that would partially demilitarize strips along the border. The zonal approach has three fundamental disadvantages; one is the creation of zones of different security; the second is the unavoidable incentive for both sides to plan how to rush through these zones and take advantage of the depletion of the other side; the third is that such a setting might create very destabilizing mobilization and break-out dynamics in case of a crisis. What really counts is to avoid attack-capable force concentrations and that can be achieved in a much more comprehensive way by overall density limits that apply to the whole region from the Atlantic to the Urals.

Complementary to these selective reductions of the offense-capable platforms in a future arms control regime, there should be restrictions on offense-capable logistics, forward-deployed bridging equipment, and ammunition stockpiles. But this is conventional wisdom and not discussed here in detail.

2. The second component of an arms control regime relates to the air forces, and it is important to understand that it is at least as crucial for conventional stability and the regime as the land forces. The most decisive bonuses for pre-emption exist in the realm of the air forces, and unless we find solutions for that problem, too, all our endeavors to increase conventional stability will be more or less futile. The basic avenue here is to reduce selectively the capability for deep, high-precision strikes. This goes exactly against traditional NATO thinking. NATO regarded its advanced strike capabilities as a crucial instrument to compensate the quantitative superiorities of the land forces of the WTO and as an instrument to diffuse attack-capable force concentrations. But these two functions disappear with the arms control regime described here. If one asks under these changed conditions which side

would take more advantage of strong air strike capabilities, we see that it obviously would be the attacking side, because it would drastically increase its bonus for pre-emption.

In a regime of conventional stability, the air forces will have a new function. Their major task is to prevent air superiority over one's own territory. For this purpose it is necessary to shift the emphasis away from penetration-capable ground attack airplanes in favor of those planes that are survivable and efficient interceptors. As to the survivability aspect it would obviously be an advantage to shift the emphasis more toward more vertical takeoff planes. They are less capable regarding payload and range but in the context described here, this is more a virtue than a disadvantage. However, the drastically reduced vulnerability of vertical takeoff planes would be a major advantage regarding the goal of minimizing pre-emptions bonuses and regarding the task that both sides possess forces that can afford it militarily, not to strike first but to wait and take over the role of the defending side. Translating these considerations into concrete numbers, one could, for example, define a regime that allows for 500 to 750 traditional ground attack airplanes and two times as many planes with clear interceptor characteristics. (Beyond doubt, definitions in this regard will be difficult and complex, but looking at the stability features of a future arms control regime, it is worthwhile to embark for the development of them.)

3. The third and last component of a regime of conventional stability is the link to the overall nuclear deterrence regime. It will no longer be necessary to compensate conventional inferiority, and thus the need for "escalation dominance" no longer exists. Especially the second point is of major importance as the idea of escalation dominance was per se incompatible with the idea of arms control. (The basic principle of arms control is equal security, and escalation dominance is per definition a one-sided feature.)

The function of the nuclear component in such a future arms control regime is (a) to deter the other side from using their nuclear forces, and (b) to provide a factor of ultimate incalculability for any war. The second function is important, too, because otherwise even a highly stable conventional regime could lead to the assumptions that war is now more "feasible" again and might again become a tolerable tool of policymaking. This must be avoided, and it can be avoided if the nuclear components of both sides, though drastically reduced in size, do not only contain the "big sticks" but also smaller yield weapons that the desperate defender might consider using without immediately launching a global Armaggeddon. The important fact in this regard is the existence of those weapons, not the questions of whether a formal no-first-use position is taken or not. What we are talking about here is existential deterrence (quite close to the French position) instead of extended deterrence.

There exists a problem with such an existential deterrence posture regarding the Federal Republic of Germany (FRG). Going from extended to existential deterrence, FRG, not having nuclear weapons on its own, might feel to a certain extent "unprotected." But this argument is not really convincing as any invasion into FRG would constitute an existential threat for France and the other European countries as well as a most fundamental violation of U.S. interests. Therefore, the transition from extended to existential deterrence should increase the net security of FRG, too.

As to the numbers, a few hundred mobile, land-based, single warhead missiles with ranges below 500 kilometers (according to INF) would be completely sufficient to fulfill the function of the future nuclear component in Europe. Therefore, one should sit down with the Soviets, better in an informal dialogue than in formal negotiations, and jointly design such a stabilizing nuclear component for Europe. I would argue that its existence would also be in the long-term interest of the Eastern side, as it stabilizes the transition processes in the conventional realm and as it—seen from an Eastern worst case analyst's point of view—would diffuse all possible future Western plans to take advantage of technological and economic superiority and, after a phase of drastic conventional cuts, reaccelerate the arms race in order to gain a position of conventional superiority. The existence of a nuclear component in Europe would reliably deny all those options. Therefore, the existence of such a minimized, stability-oriented nuclear component in Europe should be in the interest of the Soviet Union and its Eastern allies.

In addition to these three core components of the conventional stability regime, three more problems have to be solved: transition, verification, and modernization. Though we cannot go into details about the principle of how to approach these problems, they are outlined below.

1. For the transition problem, one would like to have a formula that combines two effects—drastic cuts and a continuously improved force ratio in favor of the weaker side during the process. A feasible formula for this might be that one starts by defining the final ceilings (e.g., 10,000 MBTs) and then cutting the respective surpluses in certain intervals by half, three or four times over. This means that the surplus of the WTO would initially be more than 40,000, the surplus of NATO 10,000; in the first round the WTO would have to reduce by 20,000, NATO by 5,000. In the second round, the WTO surplus would then be 20,000,

the NATO surplus, 5,000; that is, the cuts would have to be 10,000, respectively, 2,500, and so on. The advantage of this approach is that it consists of just one simple formula that is applied identically to both sides and nevertheless rapidly evens out all initial superiorities. (A positive side effect is that one can work this on each side's own data base. The side that would confess too little would not profit but would be punished. The lower the starting point is, the lower are their intermediate ceilings, that is, the more systems have to be destroyed under international control.

2. As to verification, it is important to find a fairly simple but comprehensive regime. The INF Agreement demonstrates how expensive verification procedures can become. If one would apply the same approach to the conventional realm, we might easily end in a situation where the verification costs almost exceed the saved military expenditures. But there are much cheaper solutions and one of them might be a regime of "brigade observers." These would be officers in the rank of a colonel or so, who would be permament guests of each brigade, respectively, regiment commanders of the opposing alliance. Those brigade observers would at the same time see what happens in the day-to-day practice, participate in all maneuvers, and would make sure that there are no major troop concentrations, and such. They would, therefore, provide a very reliable and comprehensive verification regime without costing much. Analogous observers should be accompanying air force units and eventually even navy units, but naval arms control will be the task for the next decade.

3. The last problem we tackle here is modernization. On the one hand, it would be unrealistic to try to stop modernization altogether. On the other hand, one must make sure that the stability features we are striving for are not undermined by destabilizing modernizations. The problem is that, per definition, one cannot know what will be invented tomorrow. Therefore the idea of a comprehensive listing of destabilizing developments is a futile approach. So what do we do?

   A fairly simple but elegant solution might be the following rule: Modernization is allowed. It just must be announced from the very beginning, and if side A introduces x new weapon systems of a specific kind, it automatically allows side B to introduce two times as much of an analogous system.

   What does this 2:1 rule mean? Modernizations that strengthen defense capabilities clearly more than offensive capabilities (i.e., decoupling factor > 2) would be advantageous. The introduction of an efficient new mine system, as described above, would be a sensible modernization. Even if the other side is allowed to introduce two times as much, there is a net profit of one's own security as (a) the other side does by this not increase its offensive capabilities, and as (b) the other side is channeling its investments away from highly offense capable components into these attrition-oriented, predominantly defensive ones.

Yet, one would think twice before introducing a new ground attack airplane or a new major battle tank. Here the decoupling factor of defensive and offensive capabilities is not so clear, and one easily might shoot oneself in the foot.

Such a modernization regime makes sure that modernization per se is allowed, but at the same time it makes sure that only those modernizations are realized that overproportionally strengthen the defender. And by providing this effect, the modernization process is tuned to the overall stability goals, that is, we are getting a highly flexible highly innovative regime but are making sure that the stability features are not undermined but improved by these changes.

I stress in conclusion that the proposals and arguments given above are definitely not the final solutions. The purpose is just to give an idea of what it would mean if we take the goal of conventional stability seriously and encourage a more conceptional, more innovative, and more courageous approach.

Chapter 2

Defensive Strategies

Jasjit Singh

Great interest has been generated in recent years in security paradigms seeking alternate approaches through change in military doctrines toward defensive strategies variously described as "nonprovocative," "defensive," or "nonoffensive defense." Although there are elements of differentiation in the doctrines, the broad concept is common to all; and for the purpose of retaining a focus, the concept as outlined by Mikhail Gorbachev may be accepted: to create "such a structure of the armed forces of a state that they would be sufficient to repulse a possible aggression but would not be sufficient for the conduct of offensive operations."[1] A related aim is to "rule out the possibility of surprise attack. The most dangerous types of offensive arms must be removed from the zone of contact."[2] Many formulations have been put forward by strategists and analysts; however, some basic questions on the concept and its credibility remain.

Historical evidence supports the view that given the time, will, and means, there is no defensive structure that cannot be penetrated by a determined aggressor, and hence the need to include an element of "offensive" capability in the defensive doctrine. Essentially, therefore, we are talking of a mix of offensive and defensive capabilities, with a dominant role for defense with the objective of achieving arms race and crisis stability.[3] Two fundamental issues must be noted before we can go into detailed examination. At the general level, historical experience leaves little to generate confidence in defensive strategies. India for nearly 5,000 years had adopted what may be broadly accepted as defensive defense strategies; and time and again wars through external aggressions were imposed on it. Yet there were four periods of exception in its history, which only prove the point—the Kanishka Empire, the Chola Empire (these two for short periods), the early Moghul Empire, and the British Indian Empire maintained imperial strategic outposts *beyond* the state boundaries and thereby ensured security against external aggression. The United States has remained secure against external aggression through basing its defenses on the farther littorals of the Atlantic and Pacific oceans. Yet is is also clear that the contemporary trends in strategic doctrines built up over the past few centuries are unlikely to provide durable security in the future, especially at affordable costs.

NATURE OF DOCTRINE

The current debate has generally focused on *military* doctrines and, consequently, emphasises the aspects of force structures, military hardware, and force postures. Even within this field, the tendency is to highlight the issues at the military operational and tactical levels, rather than the strategic.[4] The debate has remained essentially Eurocentric. The debate and discussions also revolve around what is essentially an arms control process to maintain strategic stability without any real fundamental changes—hence the need to look deeper, and wider.

Military doctrine by itself is an inadequate framework for a credible paradigm of peace and security. It is ultimately the political factors that provide the incentives and motive power for war and peace. The concept of non-offensive defense put forth by Gorbachev would have little strength or substance without the political philosophy from which it drew its strength and which was earlier spelled out in the Delhi Declaration.[5] The military concepts, regardless of the structural and doctrinal changes that may be introduced, would have little credibility if the political philosophy of "exporting revolutions" or "teaching lessons" to neighbors are not first renounced. Otherwise little faith can be placed in a "Peoples' War" concept as a strategic defensive doctrine. A nonoffensive defense doctrine, in contrast, is a viable credible proposition in the context of the Indo-Pak Simla Agreement (1972), renouncing the use of force in the resolution of disputes between the two countries.

Military doctrines are concerned with the employment of military forces and hardware to achieve the desired objectives. However, those objectives are really defined by political (and, sometimes, ideological and even religious) factors. Military force, thus, is only an instrument of politics. It provides one of the means, albeit the ultimate and most destructive one, to pursue political aims. Therefore, it is the nature of political philosophy that would ultimately define the offensive or defensive character of the employment of military force. There is undoubtedly an interactive relationship between political philosophy and military doctrine, but except in the case of wanton application of military force, it is the political philosophy that defines the aims and limits of employment of military force, and hence its doctrine and hardware. It is not that military force and hardware are not important issues, for they also have a role in shaping perceptions and political philosophies. The point is that the two must be taken together with emphasis on the political element to get a comprehensive approach. We may therefore define the first paradigm of defensive strategies as: *"Nonoffensive defense must be conceived as a politico-military doctrine for durable peace and equal security for all states."*

ROLE OF FORCE

Contemporary politico-military doctrines have been primarily guided by the Clausewitzian concept of "war (and use of military force) as an instrument of policy."[6] War as a continuation of politics by other means implicitly creates a paradigm of strategic offensive politico-military doctrine. War and use of military force can possibly serve as an instrument of policy only for the initiator of war and the attacker. For the defender, it does not really signify a policy option because the victim's choice lies between fighting back and capitulation. In contrast, the very concept of nonoffensive defense is constructed with the prime objective of war prevention—even by accident. The doctrine, therefore, would carry little credibility without renouncing the concept of war as an instrument of policy.

As it is, war is a continuation of politics by other means (that is, military force) only when in the calculations of the initiators of the war, it is likely to result in a favorable outcome in terms of a cost-benefit analysis. The subjective perceptions and calculations of the initiator become the predominant factors in shaping the employment of military force. Nations believing in war as a legitimate instrument of policy are bound to retain predominantly offensive capabilities and doctrines in their military establishment in order to maximize their favorable cost/benefit ratios; and the incentive to initiate war increases as a consequence.

As it is, the cost/benefit ratios can be relevant only in circumstances of a degree of symmetry among the ratios applicable to the two adversaries. This situation was generally applicable to the Europe of Clausewitz as much as of present times. However, this may not necessarily be operative in other parts of the world. If one adversary chooses to fight at an asymmetric plane (as the Viet Cong did in Vietnam) or is able to absorb disproportionately high costs (as Iran did in terms of casualties during the Iran-Iraq war), war no longer can serve as an instrument of policy. What is more important, the concept does not remain compatible with nonoffensive doctrines. Under these circumstances, nonoffensive strategies cannot meet their primary objective of war prevention, or the secondary aim of providing adequate defense. The problem is similarly compounded by technological asymmetries. At the same time, the role of force has been undergoing changes under the impact of technological advances and geopolitical imperatives. The process of decolonization has given rise to a large number of sovereign nation states with a high degree of nationalist fervor. This, coupled with their intrinsic insecurities and vulnerabilities, has increased the incidence of armed conflicts in the developing world. However, wars are no longer taking place for conquest of territories, and territorial boundaries have rarely been altered as a consequence of war. In contrast, increasing use of military forces as a political instrument is being made without necessarily leading to war. The two superpowers alone are estimated to have employed force without war as an instrument of policy in more than 426 incidents between 1946–82.[7] This type of use of force requires military forces with substantive offensive capabilities—rapid deployment, long-range strike, and power projection forces. The U.S. Commission on Integrated Long-Term Strategy, in fact, strongly advocates long-range surgical strikes as an important policy option.[8] Such concepts and capabilities for the use of military force are antithetical to the basic precept of nonoffensive defense. We may thus define a second paradigm: *"Nonoffensive defense, by definition, is a war prevention politico-military doctrine. The concept of war (and use of military force) as an instrument of policy is incompatible with the doctrines of nonoffensive defense."*

With the advent of the nuclear age, the concept of war as an instrument of policy has become questionable. In case of a nuclear war, no rational policy objective can be achieved because of the very high levels of destruction implicit in such a war, and the means themselves would destroy all ends. Wars involving nuclear weapon powers are bound to rapidly escalate into a nuclear exchange with inevitable results. In fact, defensive strategies may have a deleterious effect in eroding nuclear thresholds. By definition, such strategies would allow only limited offensive and counteroffensive capabilities. However, breakouts may occur due to tactical or technological surprises. Under the circumstances the losing side would have a strong incentive for early resort to nuclear weapons in an attempt to redress the balance. Conventional wars between industrialized developed countries have also lost all relevance in terms of serving any rational policy objectives. Such countries are highly vulnerable to conventional wars because of the extensive existence of nuclear power plants, chemical industries, and plastics/synthetic fibers used in all facets of societal life. Damage from conventional warheads could result in unquantifiable collateral damage effects including through pyrotoxic pollution. Intentional or unintentional attacks on chemical and nuclear plants may come early, eroding nuclear thresholds. But war perhaps may retain some of its characteristics as an instrument of policy only in conflicts involving less-developed countries. This quality will erode as developmental processes increase vulnerabilities. Yet, the Clausewitzian approach has tended to make force an integral element in the foreign policy of states, and the primacy of force provides increased incentives for using offensive force because of perceived greater payoffs in meeting policy objectives. The reality, of course, is that during this century most of the countries that initiated war in pursuit of perceived goals actually lost the wars. One may, therefore, postulate a third paradigm, which emphasises the futility of war: *"Concept of war as an instrument of policy is redundant and invalid in contemporary world."*

THE PROBLEM OF DETERRENCE

Nonoffensive defense doctrines that support military force structures that are sufficient to repulse a possible aggression but would not be sufficient for the conduct of offensive operations would appear to provide a stable deterrent paradigm. As stated earlier, however desirable, this contains strong potential for breakout, especially through technological surprise. The doctrines presuppose extensive transparency, inspection, and verification of both sides. This can be both a source of strength and a potential problem. For example, most modern weapons systems, especially defensive systems, are heavily dependent on electromagnetic spectrum for their operational efficiency. This also makes them highly vulnerable to electronic warfare by an adversary. Electronic warfare capabilities are the most difficult to verify. Higher levels of electronic warfare capabilities, otherwise not perceived as "offensive weapons," could crucially alter the defensive strategies paradigm and provide a decisive edge to the side able to use them with advantage. Mere possession of such capabilities could provide incentive for breakout from the deterrence stability.

A greater problem arises with the type of deterrence. Deterrence through denial may generally be accepted as compatible with nonoffensive doctrine, even where counteroffensive capabilities are required for denial purposes. Deterrence had historically operated only through denial until the advent of air power (and later, nuclear weapons) in the twentieth century when the additional dimension of deterrence through punishment, independent of the denial component, started operating.[9] Strategic bombing and nuclear weapons were able to target nations and their assets beyond the military forces in contact and thus directly threaten the enemy heartland to inflict very high levels of damage and destruction. This strategy of punishment forms the basis of nuclear deterrence in the contemporary world. The rationalization of this form of deterrence as "retaliation" does not alter its basic function of punishment. Retaliation can form part of both elements of deterrence: by denial and by punishment. The other aspects of deterrence through punishment apart, the fundamental truth needs to be recognized that it constitutes both an offensive doctrine and a provocative one.

Deterrence through punishment is exercised through offensive air power (strategic bombing, offensive air operations primarily against surface targets, with strike aircraft and/or missiles) and/or nuclear weapons.[10] Whereas some people may argue that this form of deterrence serves a stratetic defense purpose, the undeniable fact is that the means and methods are offensive in quality and nature. Deterrence through punishment also becomes highly provocative because it forces the target country to respond to a very high level of destructive threats to intrinsically vulnerable elements of the state. It is not relevant here as to which side initiates the process of deterrence through punishment or what the incentives are. The important point is that the strategy of reliance on strategic offensive air power and nuclear weapons provokes the potential adversary to find adequate responses, and the most "rational" response is reliance on deterrence through punishment! NATO, for example, felt completely justified in going in for a strategy of massive retaliation with *nuclear weapons* against the perceived threat of massive conventional military forces of the Warsaw Pact countries. However, the Soviet Union's acquisition of nuclear weapons provided the means for deterrence through punishment against perceived NATO threats.

The situation inevitably leads to the search for ever higher levels of capabilities for punishment; and in the absence of a complete withdrawal from this strategy option, at best a strategic balance on the basis of equivalence or parity can be achieved. This, in fact, was what happened with SALT I, which was an agreement to build *up* to agreed levels of weaponry. However, the fundamentally offensive and provocative nature of the doctrines remains, especially because of the risk of breakouts primarily through technological developments. Such strategies obviously are antithetical to the concept of defensive strategies. A fourth paradigm for defensive strategies, therefore, may be defined as: *"The concepts of deterrence through punishment are antithetical to those of nonoffensive defense."*

NUCLEAR WEAPONS

The fourth paradigm raises the question of the role of nuclear weapons, especially in defensive strategies. After the INF Treaty, progress, even if slow, is visible in respect to agreement for the reduction of strategic weapons by the United States and the Soviet Union. However, there appears little prospect for shorter range and battlefield nuclear weapons being eliminated in the foreseeable future. The pressure indeed is in the opposite direction as seen in the demands for modernization, especially by such countries as the U.K. Nuclear weapon states other than the two superpowers have clearly indicated their unwillingness to enter into the nuclear disarmament process in the foreseeable future. The debate and discussions on nonoffensive defense and reasonable sufficiency in defense, although examining the scope for restructuring and reducing conventional forces in Europe, appear to accept nuclear weapons as inevitable and necessary. Some experts have even argued that a nuclear first-use option must be retained as a hedge against disaster even if the conventional balance is improved in Europe.[11] Some of the leading advocates of NATO/WTO force reduction including short-range missiles only visualize reduction and not elimination of missiles; and other nuclear weapons such as artillery- or aircraft-delivered ones are apparently assumed to remain unaffected.[12] These may well be considered for reduction or elimination in a different framework, but the fact that nuclear weapons are considered to continue to have a role in defensive strategies in itself has an important bearing on the concept and credibility of those defensive strategies.

Nuclear weapons are both offensive and provocative in nature. There are also serious problems with tactical nuclear weapons, especially those that can be fired from dual-capable tubed artillery in the 25–30 kilometer range. Threat of a conventional breakthrough could lead to erosion of the nuclear threshold and early resort to nuclear weapons, with resultant escalation.[13] Political control and release of such weapons under the pressures of time and fog of war pose serious problems. Escalation control in the stress of combat would be even more difficult to exercise. However, as stated earlier, defensive strategies visualize only limited offensive capabilities inadequate for the conduct of offensive operations. Under the circumstances, any breakthrough in the defenses could increase the incentive for using nuclear weapons, with or without employing available offensive capabilities.

In contrast, it is significant that application of the doctrines of nonoffensive defense is being seriously considered, primarily in the European context. It can be argued that it also shows signs of some movement toward success. If it moves toward greater success in spite of the fact that the political relationship between the East and West have not reached the level necessary for the confidence required, it will be because of the availability of nuclear weapons, both at the strategic and tactical levels with both alliance systems. Thus a paradox becomes clear. Nonoffensive defense doctrines become feasible at the conventional levels, whereas nuclear weapons, essentially offensive weapons, are retained as insurance against possible failure of the defensive strategies paradigm. In reality, therefore, what is being attempted in Europe is only a partial application of defensive strategies that would reduce the levels of armaments and military expenditures. This is certainly to be welcomed. However, it cannot provide credibility to the concept of defensive strategies, because the final guarantor for peace and security would still be perceived and maintained in the shape of nuclear weapons. The paradigm really gets reversed: an overwhelming offensive capability (of mass destruction with nuclear weapons) on one plane, and strengthened defensive strategies on a different plane (of conventional weapons). The overall paradigm, therefore, no longer remains nonoffensive. The aim of nonoffensive defense, that "the most dangerous types of offensive arms must be removed from the zone of contact," cannot be fulfilled without the elimination of nuclear weapons both at the tactical and strategic levels. (The "zone of contact" in the latter case is the globe itself in view of the reach and range of such weapons.) Some experts have demanded the elimination of tactical nuclear weapons,[14] and the Soviet Union had called for the elimination of all such weapons by the end of this century. However, in the context of our examination of the concept and credibility of defensive strategies, we need to recognize the fact that nuclear weapons are representative of massive offensive capability and incompatible with the doctrine of nonoffensive defense. We may thus define a fifth paradigm that: "*Nuclear weapons are not compatible with the concept of nonoffensive defense.*"

INTERVENTIONS

If the concept of nonoffensive defense is to be examined beyond the Eurocentric framework, the harsh realities of interventions and low intensity conflicts have to be faced. It is also necessary to see the concept in the global military context because most of the 200-odd wars and armed conflicts since 1945 have taken place outside Europe. In the ultimate analysis, the credibility of the concept will really be tested in terms of its universality, especially without the protective umbrella of nuclear weapons. It may be possible to formulate credible paradigms of nonoffensive defense strategies applicable to different groups of states, with perhaps some degree of variation due to variance in environmental parameters. However, the issue of external, especially great power, interventions is crucial in its impact on the success or otherwise of defensive strategies at the global level because of its potential to rapidly alter the military balance, in offensive as well as defensive aspects. This aspect needs serious attention, especially when it is recognized that external interventions took place in three-quarters of the inter- and intrastate conflicts since World War II.

There are, of course, serious problems with assessing the nature and legitimacy of interventions because of the complexity of the paradigm of interventions (outlined in a simplified form in Fig. 2.1.) Intervention involves coercive external interference in the affairs of a population that is organized in the form of a state. Thus intervention can be for and on behalf of a state or its population, and it can be against a state or its population. It may be undertaken by a state or by nonstate actors (including military/political organizations not represented by a state, as is the case with the PLO). The principle that interventions are impermissible against a state has been gen-

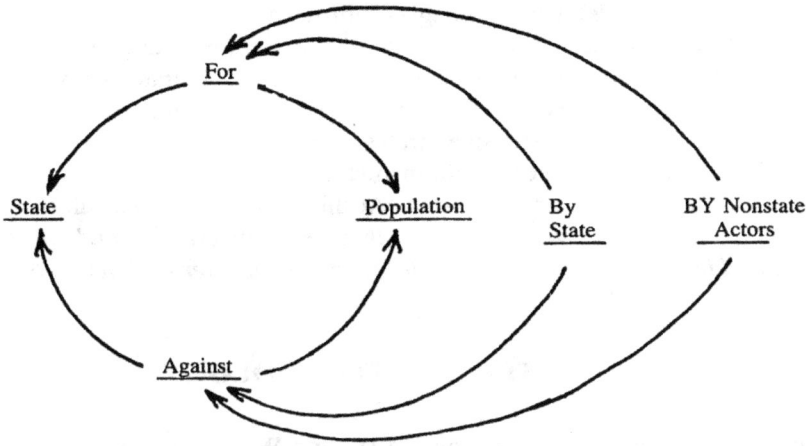

Figure 2.1. Paradigm of interventions

erally accepted and enshrined in international law,[15] and little has been said about permissibility of interventions for and on behalf of a state. However, keeping the moral issues aside, it may be stated that intervention is in general wrong because of its tendency to lead to counterintervention, escalation, and wider war. An initial intervention may give the advantage to one side in the conflict, which may not deserve the assistance. The counterintervention is then justified because it deprives the undeserving side of its advantage. The judgment on whether a side "deserves" assistance is in itself a highly subjective process superimposed by political philosophies and interests. Interventions, therefore, contain a potential for instability.

The problem, from the point of view of our present examination, is how to draw up a credible paradigm of defensive strategies in the context of interventions. The concept of defensive strategies as put forth so far fails to address the issue. At the political levels attempts have been made to define the principles and doctrines of interstate relations (e.g., the UN Charter, the concept of peaceful coexistence, nonintervention, and noninterference in the internal affairs of other states as outlined in *Panchsheel* and the Delhi Declaration of 1986 do provide the political framework). It is in the sphere of the military doctrine that the real problem arises. If a state is to build into its defensive strategies the capabilities to successfully counter external interventions by states beyond its immediate neighbors, it may be building up military capabilities that would appear to be threatening and provocative to its neighbors. Yet as long as states maintain long-range power projection capabilities, rapid deployment forces, extraregional military command structures and bases, and, above all, politico-military doctrines supporting military interventions, defensive strategies will need to take these aspects into account. In the absence of the rejection of interventionist doctrines and capabilities, the concept of defensive and nonprovocative strategies would require at least building up capabilities to raise the costs of interventions. Even then, it is not certain if defensive strategies would possess adequate credibility. The answer really lies in eliminating offensive capabilities supporting interventionists doctrines and strategies. At this stage, a sixth paradigm needs to be outlined: *"Defensive strategies, if they are to be credible and effective, need to address at the global level, the problems of external interventions."*

STABILITY AND TECHNOLOGY

Strategies of European defense, by NATO and WTO, have generally been based on a combination of two fundamental means to achieve military stability: through a balance of conventional forces and a threat of nuclear escalation. The concept that parity or balance in military/nuclear capabilities would provide symmetry and hence stability has also led to the massive military buildup in Europe. Nonoffensive defense concepts seek a change in this paradigm. However, as long as it remains limited in scope as apparent in present discussions, it would only achieve arms control and, to some extent, improved mutual confidence. The problem lies in searching for peace and security in terms of military balance and stability without restructuring political relationships.

Stability in terms of military balance is a function of the removal of asymmetry of military capabilities. However, there are different dimensions of stability as may be seen in Table 2.1. It is obvious that technology would play a crucial role in achieving stability or upsetting it.

Thus stability may be sought at different levels of armaments, and it may be sought through a spectrum of force structures depending on the military doctrines adopted. What NATO/WTO have sought so far has been a high level of armaments (qualitatively and quantitatively) and offense-dominated doctrines (although WTO denies that there is role for nuclear deterrence in

Table 2.1.
Military Stability Dimensions

Level of Armament	Stability (through balance)	Force Structures
High		Offense dominated
Medium		Offense-defense mix
Low		Defense dominated

its calculus). Defensive strategies would aim to shift the level of armaments as far down the line as possible compatible with a defense-dominated force structure. What this really means is an arms control and reduction process, but it does not necessarily imply greater security. Lower levels of armaments have not necessarily meant lower probability of war. Every previous war was fought at lower levels of armaments. And serious doubts exist as to the credibility of defense-dominated force structures, especially in the context of technological imperatives.

Technology certainly can be exploited to bolster defensive capabilities. However, reliance on technology that has not fully matured can have disastrous consequences. At the same time, the best of technology and manpower is still susceptible to errors of judgment, as indeed happened when the Iranian airliner was shot down, killing 269 people, by air defense missiles fired by the Aegis class guided missile cruiser, the USS *Vincennes*, in the Persian Gulf. Transparency and verification can ensure that force structures agreed upon are not altered, at least to the extent that they are verifiable. We have noted the potential of electronic warfare capabilities as a destabilizing factor because of the peculiar parameters of war in the fourth dimension.[16] However, R&D in military technology is an ongoing process and also receives support from civilian R&D programs through interactive processes. Most countries in general, and the United States in particular, place great store by technological superiority as the means of providing the critical edge of advantage in military confrontations. Military R&D programs remain under heavy wraps of security. The potential for technological breakthroughs having a significant impact on the paradigms of defensive strategies will remain as long as military R&D programs remain shrouded in secrecy. In fact, defensive strategies may actually boost efforts to find technological means to gain advantages, which, given the nature of the new paradigm of lower level of armaments, limited counteroffensive capabilities, and an overall defensive posture, could be crucial. It must be noted that it is not merely technological capability but the concept of its operationalization that is important. And the way technological capabilities would be operationalized by one side may not be fully appreciated by the other. The most vivid example, perhaps, was the Israeli military operations in the Beqa'a valley in June 1982 when it dramatically defeated the Syrians, who lost 86 fighters in the air and 400 tanks and 24 SAM batteries on the ground in less than 36 hours. All the systems—E-2C airborne early warning aircraft, RC 135 electronic intelligence aircraft, remotely piloted vehicles and their sensors, and so on—were known to be in Israel's arsenal for some time. The general approach to strategy by the Israeli Air Force was also known to discerning military observers. Yet the masterly coordination of the air strategy, which proved decisive, came as a surprise not only to the Syrians, but to most of the world.

Technology is making modern battlefields increasingly transparent through reconnaissance, surveillance, and intelligence-gathering sensors. This in itself would reduce the scope for tactical surprise and *increase the premium on technological surprise*. Defensive strategies certainly would be more sensitive to technological surprise; (Syrian armed forces in the Beqa'a valley conflict were essentially in a defensive posture) and R&D contains the potential to provide such surprises. We may thus postulate a seventh paradigm for defensive strategies: *"Nonoffensive defense requires greater transparency and effective management or research and development activities to eliminate the potential of destabilizing technological breakouts."*

There is another aspect of technology that deserves note for its impact on defensive strategies. New and emerging technologies are moving toward providing a completely different range of weapons systems that would revolutionize warfare in the twenty-first century. In fact, one cannot help taking a slightly cynical view that the industrialized states, especially the superpowers, are moving toward reducing the levels of armaments presented in the attractive packaging of defensive strategies, only to release resources for investment into these new and emerging technologies—the space plane and hypervelocity delivery platforms, directed energy and new kinetic energy weapons systems, and a range of highly capable reconnaissance, surveillance, and target acquisition capabilities. The combination of these capabilities, besides initiating a qualitatively new arms race, would really be moving toward tremendously increased offensive potential against which adequate defensive systems may not be available for decades after the offensive systems are introduced.

The new capabilities for warfighting would also mark a shift back from mass destruction capabilities that has dominated the world during the twentieth century, toward discrete and selective destructive power in the future. In the absence of credible defensive systems, the security paradigms are likely to reinforce the concept of deterrence through retaliation by offensive means using similar technologies, or the present conventional and nuclear capabilities. This would mean continuation of the offense-dominated strategic stability, and defensive strategies being discussed today would start becoming redundant by the time they can actually be constructed. Although the new arms race will reach maturity at a future date, and in the interim many uncertainties remain, defensive strategies will need to take this aspect into account if they are to remain durable. We may, therefore, define an eighth paradigm:

> Emerging technologies supporting an incipient new arms race will tilt the balance heavily in favour of strategic offensive. Nonoffensive defense doctrine requires more effective control and management of the emerging arms race so as to avoid the risk of destabilization of the defensive strategies.

WAR AND SOCIETY

The debate on defensive strategies appears to focus on the defense and security of the *state*. However, there is a serious need to pay attention to the defense and security of *society* (and the environment) from war. This is because the classical paradigm of targeting priorities of war, valid up to the time of Clausewitz, has been reversed in the past two centuries. From the primary objective of war being destruction of the enemy *military power*, followed by conquest of the enemy *country*, and then the subjugation of the *will* of the enemy nation as defined by Clausewitz, the process of industrialization and totalitarianization of war has created an altered linear paradigm that must be addressed by the concepts of defensive strategies.

Society has always been involved in war preparedness and contributing the means and manpower for wars. Developments arising out of the Industrial Revolution and the French Revolution, however, resulted in society itself becoming inclusive to war and warfighting. Even in prerevolutionary eighteenth-century Europe, assessments were that "At the present day war is carried on by regular armies; the people, the peasantry, the townfolk take no part in it and as a rule have nothing to fear from the sword of the enemy."[17] However, rapid changes took place toward the turn of the eighteenth/nineteenth centuries; about which Stendhal was to remark, "What a change from 1785 to 1824! In two thousand years of recorded history, so sharp a revolution in the customs, ideas, and beliefs has never occurred before."[18]

The totalitarian nature of war in modern times started with the French Revolution and Napoleonic Wars. The American Civil War reinforced the model when mass armies were pitted against each other in years of battle, with the societies of North and South mobilized behind them. (The industrialization of war also had its major impact during this conflict.) World War I saw the model really come into its own with the emergence of political economy of war on one side and expansion of military technological means to bring society totally within the fold of war on the other. As Ludendorff was to write, "The nature of totalitarian warfare literally demands the entire strength of the nation, *since such a war is directed against it.*"[19] The "total war" model requires total mobilization of the people to provide political and ideological support for economic and military achievements. Because this becomes true on both sides, the strategy of each adversary then inevitably has to be based on attacking the *people as well as the military forces* of the other. This linear approach to the relationship between society and war demanded doctrinal incorporation of society in war. Concepts of revolutionary doctrine emphasizing the societal struggle (as against narrow professional military activity) only tended to reinforce the incorporation of society into warfare. Society and people inevitably because participants in warfare, and given the emergence of strategic thought based on the primacy of offensive military force, targeting and subjugation of people's will moved up in the hierarchy of war aims. Advances in technology made it increasingly possible to target the enemy nation's society (and its will) directly without necessarily having to destroy the enemy's military power or conquer the enemy's country.

At the fundamental level, the defensive strategies must aim for equal security for all—at the level of the individual as well as that of the state. This would require reshaping of contemporary politico-military doctrines to provide for political, military, and technological paradigms supporting this objective, and a ninth paradigm needs to be written in: *"The most important and pressing need is to alter the society-war-society linear paradigm so as to make society exclusive to war. Nonoffensive defense should remain a purely counterforce doctrine."*

This has become all the more necessary since modern societies, especially those at higher levels of industrial development, are increasingly more vulnerable to armed conflict across its full spectrum, from nuclear and conventional war to low intensity conflict and terrorism. Today's Europe, with all its nuclear power stations, chemical plants, and extensive use of plastics and synthetic fibers, is far more vulnerable even to conventional bombing than the Europe of 1939–45. Even in World War II, the conventional bombing raids on Tokyo on the night of March 10–11, 1945, caused greater damage than the atomic bombing of Nagasaki five months later. The developing countries are increasingly becoming vulnerable to armed conflict as they move up the developmental ladder. In some respects, their society is extremely vulnerable to conventional strikes against high value economic targets and population centers because of their greater dependency on the former, and greater density/less protection in the latter. Nuclear war, of course, would encompass society well beyond national frontiers.

Society in general is not an active party to the conflicts between states, and killing of the innocent being a grave existential evil seems to be accepted by all sides. In contrast, it may well be argued that war and armed conflict is a *normal* phenomenon, although a great deal can be done to reduce its incidence. Meanwhile, it is imperative that society as such be insulated from the process of warfighting and its effects. Given the trends arising out of industrialization of war, this would not be easy. However, it is necessary to explore the directions that the search must take, and this will need to emerge out of changes in politico-military doctrines.

Perhaps the most important changes needed in the politico-military doctrines rest on two requirements: (1) emphasise the needs for limiting/restraining conflicts, de-emphasise the offensive component in the doctrines, and (2) eliminate capabilities for large-scale mass destruction. Limited warfare by itself tends to limit the involvement of society and populations into war. Society can be substantially kept outside the "battlefield" if wars are restricted to military forces and installations only. De-emphasising the offensive, at least at the strategic level, and elimination of mass destruction capabilities are vital if society is to be kept insulated from the scourge of war. A shift back toward discrete and discriminate means of warfare has become vital. International agreements and legal regimes will need strengthening toward this end. However, the most important step necessary is the change in ideas and attitudes toward the function of society as a target in war.

POLITICAL PARAMETERS

As stated earlier, political philosophies and relationships are prerequisites for the credibility and effectiveness of defensive strategies. The main reason why a Eurocentric approach to formulation of defensive strategies has emerged is that there is no conceivable cause for a military conflict, and political relations have been improving between the Eastern and Western blocs. Mututal threat perceptions have undergone significant changes in the recent past. For example, recent polls showed that only 4 percent of the British public now sees the Soviet Union as a serious threat to British security, and 80 percent of the West Germans now believe that there is no longer a military threat from the East.[20] Force reduction under such political climate becomes easier. This is also reinforced by the fact that Sino-Soviet political relations started improving well before force reductions could be initiated. Thus the key to the successful application of the defensive strategies would appear to lie in political philosophies and relationships. We may define an important paradigm (the tenth) as: *"Nonoffensive defense is conditional and sensitive to political philosophies, objectives, and relationships."*

Against this background it becomes clear that political concurrence by potential adversaries is necessary for the credibility of defensive strategies. War is an interaction of two forces propelled by two sets of perceptions, intentions, and capabilities. Contemporary politico-military doctrines rely on creating insecurity and vulnerabilities in the opponent's mind and state. A shift back to a more positive paradigm requires *mutual* understanding and acceptance of military balance and stability in the new framework. The element of mutual reciprocity therefore becomes crucial. Defensive strategies cannot truly be implemented unilaterally. Potential adversaries, whether they are immediate neighbors or distant states, will need to accept the concept and political, military, and technological restructuring and constraints. The eleventh paradigm becomes another important one: *"Nonoffensive defense doctrine can be successfully applied only through mutual reciprocity and universal acceptance of the concept."*

This in itself raises an interesting issue. If states accept the concept of nonoffensive defense, the logical process would be a progressive movement toward complete and general disarmament. Any other approach would leave it as another partial arms control measure without any fundamental change from the contemporary and dangerously unstable doctrines. Once states mutually concede that war is not an instrument of policy, that they will not maintain offensive capabilities that are threatening or provocative, the real need for defensive capabilities also disappears. Defensive strategies by themselves are incomplete paradigms and fail to meet the criteria of credibility unless a total and comprehensive approach is adopted for their universalization. The ultimate goal and objective of defensive strategies must be clearly kept in mind when examining specific elements. In this context, the twelveth paradigm may be defined: *"Nonoffensive defense is a transitional doctrine to complete and general disarmament."*

Wars are made in the minds of men, and it is in the minds of men that defenses of peace must be constructed first. Humanity stands at the threshold of a historical dysjunction of increasing military standoff resulting in strategic defensive environment and strategic offensive doctrines and capabilities constructed on the calculus of generating fear and threat in the opponent; a dysjunction between the political futility of military power and doctrines built on the primacy of offensive military force. Defensive strategies and doctrines hold out the promise of harmonizing the needs of states and societies for mutual benefits. They also can be credible only if mutually accepted and implemented. However, a great deal of groundwork—at the conceptual, doctrinal, and operational levels—has to be done before these concepts move close to being credible realities.

Notes to Chapter 2

1. M. S. Gorbachev, "The Reality and Guarantees of a Secure World" *Pravda*, September 17, 1987. The change in WTO military doctrine was formally adopted at the Berlin meeting of the Political Consultative Committee of the Warsaw Treaty Member States, May 28/29, 1987.

2. *Pravda*, February 14, 1987.

3. See Alvin M. Saperstein, "Primer on Non-Provocative Defence," *Arms Control*, Vol 9, No. 1 (May 1988).

4. See the Special Section on Non-Offensive Defence, *Bulletin of the Atomic Scientists* (September 1988).

5. Delhi Declaration, November 26, 1986, issued by Prime Minister Rajiv Gandhi and General Secretary M. Gorbachev on principles of a nuclear weapon free and nonviolent world.

6. Jasjit Singh, "Evolution of Politico-Military Doctrines," in Air Cmde Jasjit

Singh and Dr. V. Vekaric (eds): *Non-Provocative Defence: The Search for Equal Security* (New Delhi Lancer International, 1989).

7. Barry Blechman, Stephen Kaplan, et al., *Force Without War: US Armed Forces as a Political Instrument* (Washington, DC: Brookings Institution, 1978); Philip Zelikow, "Force Without War, 1975–82," *The Journal of Strategic Studies* (London), Vol. 7, No. 1 (March 1984); Stephen Kaplan, *Mailed Fist, Velvet Glove: Soviet Armed Forces as a Political Instrument* (Washington, DC: Brookings Institution, 1979).

8. Fred Ikle, Albert Wohlstetter, et al., "Discriminate Deterrence," The Report of the Commission on Integrated Long-Term Strategy (January 1988).

9. Johan Jorgen Holst, "Denial and Punishment: Straddling the Horns of NATO's Dilemma," *Adelphi Papers*, No. 206, (London: IISS, 1986); Jasjit Singh, "Conventional Deterrence in the 1990s," *Indian Defence Review* (January 1986).

10. Nuclear weapons themselves, especially in a strategic role, require the use of offensive air power for delivery—by aircraft, ballistic missiles, cruise missiles, etc. Only some category of tactical nuclear weapons are delivered by other means, as in the case of nuclear artillery.

11. Richard K. Betts, "Conventional Deterrence: Predictive Uncertainty and Policy Confidence," *World Politics*, Vol. 37, No. 2 (January 1985).

12. Jonathan Dean, "How to Reduce NATO and Warsaw Pact Forces," *Survival* (March/April 1989).

13. Jasjit Singh, "Eroding Thresholds," in K. Subrahmanyam (ed): *Nuclear Proliferation and International Security* (New Delhi Lancer International, 1985).

14. Andreas von Bulow, "Defensive Entanglement: An Alternative Strategty for NATO," in Andrew J. Pierre (ed): *The Conventional Defence of Europe: New Technologies and New Strategies* (New York Council on Foreign Relations, 1986).

15. For example, see UN Declaration on Principles of International Law, 1970.

16. Jasjit Singh, *Air Power in Modern Warfare* (New Delhi Lancer International, 1985), see chapter on "War in the Fourth Dimension."

17. Vattel in *Law of Nations*, quoted by Sir Basil Liddel Hart, *The Sword and the Pen* (London: Book Club Associates, 1978).

18. Stendhal, *The Charterhouse of Parma*, trans. C. K. Scott Moncrieff, (London: Chatto and Windus, 1962).

19. Eric Lundendorff in *The Nation at War*, cited by Hart, n. 17.

20. Jack Beatty, "The Exorbitant Anachronism," *The Atlantic Monthly* (June 1989).

Part II

Defensive Strategies in NATO/WTO Relations

Chapter 3

Prospects for Arms Control

Andrzej Karkoszka

Both NATO and the Warsaw Treaty states have declared their strategies and their underlying military concepts to be "defensive." As a yardstick and ultimate proof of this defensiveness, both sides put forward their stated lack of intention of launching a war against the other side. The enormous military potentials of both alliances were justified as being necessary to compensate for the other's offensive capabilities and not by any desire to start any unprovoked offensive operations on their own. The whole argumentation purposefully overlooked the fact that both have the capability, in the technical sense, of launching offensive operations. In the view of the WTO, what should matter was that states had benign intentions. The actual military strength on which these intentions were based ought to have been seen as a secondary matter. NATO has long ignored intentions, claiming that only capabilities should count in drawing up any threat assessment. That is how the logic of the past four decades of relations between the two alliances might be characterized.

This logic has been entrenched by mass propaganda, which has constantly depicted one's own intentions as beyond question and the other side's intentions as dubious or entirely false. This outlook also played down one's own military capabilities, at the same time exaggerating those of the other side. In the context of the East-West confrontation, this political and propaganda machinery has been largely self-sustaining and has played an important part in generating the Cold War.

This situation has long been illogical, not only on the account of its own inner contents. It also helped to obscure the fact that the "defensive"—read, nonaggressive—intention made virtue of a necessity: the exclusion of the possibility of large-scale offensive operations from any rational military options in Europe, which was caused by the existence of the strategic, theater, and tactical nuclear arsenals.

It may be said that both sides stand equally accused of professing this continuous discrepancy between their declared intentions and their actual military postures. Neither the large tank armies that the Warsaw Treaty states deployed near the dividing line of Europe nor the intermittent NATO drive for military/technological superiority could be taken as compatible with a truly defensive posture.

It may be assumed that the vitality and longevity of this abnormal discrepancy between the declared intentions and the actual military practice have been made possible by the existence of certain intra-alliance or internal national political motivations. Only by addressing these points can one attempt to explain the difficulties involved in stemming the mindless inertia of the East-West military confrontation.

The critical assessment of the declaratory aspect of the NATO and Warsaw Treaty military posture should not be taken as belittling the importance of a proper understanding of political intentions and declarations connected with a state's military policy. In some cases it is the intentions that are more decisive than the actual military capabilities, as witnessed by several instances or wars waged by apparently weaker states against theoretically stronger ones. This was the case with the Hitler's Germany against the coalition of European states, and of Israel against the Arab states. This point addresses the necessity of merely matching the declared military policy with the actual military posture. In this case, it may be concluded that the past and present potentials of the two alliances in Europe are not compatible with truly defensive strategies. Their declarations were until recently lacking credibility, raising suspicions and uncertainties. To be credible, the military forces of the two alliances would have to be, first, much smaller and, second, would have to undergo substantial organizational, technical, and geographical restructuring, leading to the elimination of the ability to launch surprise offensive operations while still being capable of assuring the ability to defend one's own national territory.

FEASIBILITY OF A TRULY DEFENSIVE POSTURE

It is difficult to be optimistic about the feasibility of a system called "mutual defensive dominance" if one realizes the existing gap between the conceivable size and characteristics of forces built upon such a theoretical concept, and those actually in existence today in Europe. The gap is so vast that it may even be seen as counterproductive to posit this theoretical concept as an immediate political goal, so uncertain is the path from here to there, so many are the unpredictable outcomes of various steps, and so dynamic are the political and technological factors that constantly push the developments in the military domain away from the goal and back to the old arms race habits. However, the situation sould be seen as less dramatic if one accepts as a modus operandi a gradual process in which the two alliances undertake continuous (though carried out in stages) eradication of the most obvious offensive features of their respective military apparatus. This approach avoids the issue of which particular theoretical model of "true defensiveness" this whole process should lead to. After all, there are many such models already available and many will undoubtedly be designed in the future. All of them help to develop a theory of defensiveness applied to a given military context, whereas the real political process, intellectually facilitated by this theory, would unfold according to its own rules of the game. Thus, instead of talking about the feasibility of the ultimate theoretical *models* of "defensive posture," we are talking about the feasibility of the *process* leading toward such a concept.

Aware of the complexity of such a process and taking into account the experiences from past negotiations on arms limitation, the following four general prerequisites may be assumed as being indispensable for making the process practicable: it has to be flexible in methods; it has to develop gradually; it has to be paralleled with the appropriate developments in the general political environment between East and West; and it has to be comprehensive in scope.

THE IMPORTANCE OF THE GENERAL POLITICAL ENVIRONMENT

There is no doubt that the influence of the political environment is the most decisive among all the factors bearing on the success or failure of arms control negotiations. Regardless of what formal excuses were given in the past to explain the difficulties in negotiations or even their outright failure, the true reason was the lack of political will of the negotiating parties.

Although the linkage between the political atmosphere among the negotiating states on the one hand, and the results of their interactions on the other, is fairly strong, there were, of course, instances when, despite an overall animosity, states were still able to arrive at arms control agreements. In some instances, these agreements helped to stabilize or improve political relations between the two alliances for some time, or they made it possible to bring under control military developments that otherwise would have posed a great danger to the security interests of the states concerned. They were not, however, far-reaching and did not bring about fundamental improvements in the political climate. In the case of the process of transforming existing military forces into ones establishing "mutual defensive dominance," the scope, depth, and duration of changes to be executed by states in their arsenals and structures of forces would by far exceed anything achieved so far in arms limitation, and thus they could not be possible without a deep and long-lasting parallel improvement in the East-West political climate.

The existing political environment in Europe continued to evolve visibly toward substantial improvement—from the Stockholm Final Document, through the INF treaty, to a successful opening of the Vienna CFE talks. In addition, there have been several other more or less successful meetings connected with the CSCE process. The perceptible relaxation in East-West relations may be further strengthened as a consequence of the political reforms now underway in several of the East European states, including the Soviet Union. If these European processes would continue, then, for the first time in the postwar history of the continent and of East-West relations, political conditions would clearly be conducive to the extensive transformation of military relations between European states.

THE FLEXIBILITY AND GRADUALITY OF THE ARMS CONTROL PROCESS

The experience of recent years of arms limitations efforts in Europe show clearly that once the political will is established to reach a particular end, there is no restriction on the methods that can be used to lead to a desired outcome. Apart from the classical negotiating process, several important steps were undertaken on a unilateral basis, some of which included the possibility of multilateral verification. Although it has not yet happened, one cannot preclude a possibility of some reciprocity to take place if further unilateral measures were to be undertaken. In Vienna, the negotiating parties are considering the possibility of establishing a verification mechanism—in the form of a data validation scheme—before any agreement on reductions is in sight. This would be an heretofore unheard of consensus. Several ways, including various package deals involving quite different weapons systems, are under consideration to help eradicate the existing asymmetries in military potentials. Flexibility of this kind in finding the most effective methods in furthering the military rapprochement in Europe acts as clear proof of the political willingness of all parties concerned to achieve progress and facilitates the gradual change of the negative perceptions of opponents.

The gradualist approach to the process of military transformation in Europe should be understood not only as its division into subsequent stages and periods but also as its evolution from simple to more complex measures. Straightforward numerical reductions and common quantitative ceilings, for example, cannot achieve "mutual defensive dominance." Even deep cuts within military forces and arsenals cannot transform them into credibly defensive, nonthreatening potentials whereas they might be, in given circumstances, detrimental to the overall military stability. To achieve this end, several complex, expensive, and sometimes painful steps will be required. Those more advanced measures cannot, however, be considered until the NATO and the Warsaw Treaty arsenals are diminished from their present dimensions and until the negotiating parties establish a sound basis for mutual confidence, which is within reach in Vienna.

A NEED FOR COMPREHENSIVENESS

There are several aspects of a state's military posture that bear on its ability to wage offensive as well as defensive operations as well as on the opponent's perception of the other state's "defensiveness." The only way to establish a credibly defensive posture seems to be to undertake an all-encompassing transformation in the doctrinal, organization, and technical aspects of a state's military potential. Among the components of a conceptual comprehensive program leading to a "mutual defensive dominance" are:

- Reduction of the quantitative levels of forces and arsenals.
- The defensive restructuring of these forces and of a state's entire military infrastructure.
- Adoption and implementation of the defensive military doctrine.
- Reduction of nuclear forces and the de-emphasis of their role in both declaratory and operational aspects of military postures.
- A range of measures that enhance openness and transparency in states' military policies, including various confidence-building measures, constraints, and verification mechanisms.
- Subordination of the weapons acquisition systems of states to the requirement of defensive posture.

The reductions of quantitative levels of forces and weapons may, but need not, result in leveling out the numerical asymmetries between the states engaged in the process. The quantitative aspect of arms limitation is the most visible, though not necessarily the most vital element of transformation of forces from the offensively capable to those that are defensively disposed. Being visible and more easily negotiable than other types of measures, reductions tend to draw the attention of politicians, negotiators, and the general public alike and are usually seen as a step that opens the way for subsequent, more ambitious undertakings. Closely connected with the quantitative side of the arms limitation process is the concept of parity. It is often and mistakenly believed that numerical parity is a valuable goal in limitation efforts oriented at strengthening military stability. However, it should be stressed that the most valuable aspect of numerical parity of different military potential is its political dimension and not stability itself, which depends on a number of qualitative rather than numerical factors.

The notion of the restructuring of forces and of military infrastructures covers several possible specific measures: from changes in the organization of units and changes in weapon compositions in the organizational elements of forces, to changes in their spatial distribution (deployment). Similarly, specific parts of the military infrastructure, and especially military logistics and operational support assets (e.g., bridging equipment), have a particular influence on the offensive or defensive character of a military organization. It is obvious that a reorganization of forces and military infrastructures would be both complex and expensive, especially the one so deep and comprehensive as necessitated by a program of defensive transformations in Europe.

The adoption of defensive military doctrines and unequivocally defensive strategic concepts—as they are understood by analysts in the WTO—would create a conceptual background for all other measures in the technical and organizational domains of military preparations on strategic, operational, and tactical levels. By being introduced into the officer's educational process, by shaping the training routines of forces and their exercises, the new defensive operational thought accepted by both NATO and WTO could become solid proof of the credibility of the new, defensive military postures, just as much as the far-reaching measures of a more material character.

The function of openness, confidence building, and voluntary constraints in the framework of defensive military posture does not need to be elaborated. Together with an appropriate verification of the implementation of agreed-upon treaty obligations, all of these measures are the best guarantee against any rapid or surreptitious reversal of the transformation process and against any strategic as well as tactical surprises that could endanger the security of other states.

Requirements for the reduction of nuclear arsenals and for putting the weapons' acquisition systems of states and alliances under some form of restraint seem to be the most troublesome yet, at the same time indispensable for the success of defensive transformations.

Nuclear weapons are the most destructive and, once used, the most offensive of weapons. Their existence may thus be seen as contradicting the very essence of a defensive posture. They also have a strong impact on the perceptions of security, impeding relaxation in international relations. In contrast, they are treated by many states in the West as a cornerstone of their security, indispensable even in times of "defensive dominance" as a hedge against any surprise. Hopefully, the progress in military transformations toward conventional "mutual defensive dominance" will render possible the reduction of nuclear arsenals down to a minimum, finite number, the limiting of their geographic distribution, and a slowdown in the pace of their modernization.

The issue of weapons acquisition is germane to the two highly controversial areas of military policy of states engaged in defensive transformations, namely, to the problem of technical modernization of forces and to the so-called transarmament from a predominantly offensive to clearly defensive composition of the respective arsenals. In both areas, dynamic and powerful military, political, technological, and economic factors are shaping states' attitudes. Because the weapons acquisition processes in both alliances possess their own dynamics, they are highly insensitive—especially in the Western market economies—to dictates stemming from any international arms limitation agreements. Furthermore, because the advancement of military technology is to a large extent obscured from the public eye and its consequences are not easily comprehended by either public or political elites, the demand to put it under control is very difficult to sustain. Moreover, it may be assumed that as the military forces of states in Europe are gradually reduced, the value of their technological qualities will grow in importance.

The technological progress of military arsenals cannot be forestalled altogether. But if it must proceed, it should not be guided exclusively by, on the one hand, the budgetary constraints and, on the other, by a demand for the utmost operational efficiency, by uncontrolled group interests, by the inertia of "technological sweetness," or by pure mercantile competition. Otherwise, the natural drive for technological improvements of arsenals may pose a threat to the whole concept of military defensiveness and to the durability of the process of its practical application in the East-West context.

THE CONCEPT OF DEFENSIVE DOMINANCE AND THE ALLIANCES

The feasibility of far-reaching changes in East-West military relations, from the present confrontation of highly offense capable armies to "mutual defensive dominance," can also be assessed on the basis of the existing policies of the two alliances. Such an assessment must be treated, however, as only temporary, as attitudes of the allied states may evolve substantially over time.

The activism of the Warsaw Treaty states in the area of arms limitations, the fundamental turnabout in their position regarding intrusive methods of verification, and their growing openness about their military policies seem to indicate a willingness to undertake even more profound disarmament measures than those debated in Vienna. It is visible that this positive attitude is part and parcel of all of the political changes taking place in the East and of the obvious desire to release the hard-pressed economies of the East European states from their military burdens. It is worth recalling that the Warsaw Treaty states officially committed themselves to shape their military effort according to a concept of "defensive sufficiency" and to a strictly defensive doctrine. Several concrete proposals, forwarded in recent years by these states individually or as a group, make direct reference to the concepts of "alternative defense." Taken together, the declarations and the actions of the Warsaw Treaty states seem to bode well for the feasibility of far-reaching arms limitation measures.

However, a future position of the socialist states regarding deeper reductions in their armed forces and their restructuring may not be as positive as it seems at present. The redeployment of units and logistics may prove to be expensive, withdrawal of stationed forces may become politically sensitive, large reductions in military personnel, especially the officers, may create social and economic problems, and the military may start to oppose a further degradation of their political and social positions. Moreover, deep reductions may raise the prospect of the West's technological superiority being amplified in the framework of numerical equality. Seeing their conventional forces weakened in relation to the West, the socialist states may modify their present position desirous of elimination or, at least, limitation of nuclear arsenals, partially mimicking the typical Western attitude in this regard. Finally, the reforms and political alterations taking place in a number of the Warsaw Treaty states may negatively influence the cohesiveness of the alliance, thereby damaging its resolve, or its ability, to continue the arms limitation efforts.

If the official reactions are to be taken as guidance on NATO's position, the feasibility of a far-reaching transformation in that alliance's military forces would seem very doubtful. NATO is clearly unsympathetic to any change that might undermine the presently functioning alliance machinery, disregarding the positive impact such a change might have on the East-West relations. Despite the constant wailing about the threat emanating from the East and despite all the burdens of defense expenditures, NATO seems quite satisfied with its posture and its strategic concepts, especially now when they are to obtain the quantitative parity with the WTO. The prevailing opinion up to now considered any reductions of forces below 10–15 percent of the existing levels as an anathema to its security policy, regardless of the fact that to match this level, the Warsaw Treaty would have to reduce its forces accordingly. Only recently does NATO seem to realize the seriousness of the East's willingness to shape an entirely new military order in Europe. The negative attitude may thus, hopefully, not last. Similarly, there seems to be a mellowing of NATO's stance on nuclear weapons, in the sense that its latest official position does not preclude any options for the future, including arms control options. This does not mean, however, that any fundamental change in NATO nuclear doctrine and deployments is in sight.

The strongest opposition to the concepts of purely defensive posture stems in NATO from the belief that their implementation would undermine the established patterns of force-to-space ratios in the central region, patterns dictated by the strategy of forward defense. Likewise, the concepts would throw doubt on the wisdom of the preparation for deep-strike operations. There is no doubt that deep reductions may necessitate painful decisions, especially by NATO: transforming a strong forward-deployed, linear defense into a mobile area defense supported by fortified strong points. Obviously, such a shift would cause a redefinition of the burden-sharing within the alliance and the reassignment of roles played in alliance defenses. On both sides assurances could be demanded against any chance of a surprise attack, which would give a bonus to the attacker, as well as against the prospect of a race in mobilization in time of crisis. Moreover, further reaching openness and transparency in the East-West military relations could be required.

All these considerations point to the existence of propitious political conditions justifying the analysts' thinking about a future dramatic change in the attitudes of both alliances. These conditions are not yet matched by any purposeful or decisive mutual undertaking leading to an alteration of the military situation in Europe. One may conclude that the process leading to the establishment of a new regional military system as an alternative to the "armed peace" based on the offensive capabilities on either side has not yet begun. We are not even close to embarking upon such a process. The prospective agreement in Vienna will probably establish common ceilings for the six elements of the NATO and Warsaw military postures, to be implemented over the next years. But there is no visible inclination that the negotiating parties are ready to include in the agreement any commitment for the subsequent reductions and restructurization. And without such a commitment, the common ceilings will not bring what Europe is yearning for: stability strong enough to withstand pressure even at the time of a deep political crisis.

Chapter 4

Military Restructuring and the Challenge of Europe

Gwyn Prins

This chapter sets the present time in East/West relations in its historical and political context. It argues three main points. The first is that for the first time since the era of total war opened in the 1860s, we now have an opportunity to create a new system of international security that has not required the spur of war to bring it about. This is new, unpredicted, and of the highest importance. Second, it observes that in Europe, the debate on how to handle the USSR and its allies is now over, bar the shouting. The West German view has prevailed, both in NATO and in the EC, where a Franco-German axis will determine the future orientation of the drive toward political integration. The British have finally lost the Battle of Waterloo. The third point, underlying the other two, is that in the acquisition of power and influence, the possession of concentrated military force is a sharply depreciating asset.

It is then explained why the Vienna Conventional Forces in Europe (CFE) talks are potentially the most important set of East/West negotiations since 1945. At present, the NATO countries display deep ambivalence about how to proceed; 1990–91 is thought likely to be the decisive period in which these talks may influence the larger process of European transformation: It is High Noon. The immediate problem for NATO in CFE is seen from NATO's official perspective to have three corners. It is a triangle that will not "compute."

1. How to manage cuts that are probably irresistible and are now likely to be considerably more than the cosmetic adjustments originally expected. "Manage" means both to think about and to act upon. "Irresistible" means that they are driven by more than arms control.

2. How to make such cuts without damaging the present formal NATO doctrine. "Damage" means a public admission or undeniable public exposure of inadequacy. In other words, how to confront a new situation without any fundamental change of perspective: without new thinking.
3. How, at the same time, given that the present formal doctrine may not do: (a) to draw benefit from new thinking, which may help to do so while always preserving objective; simultaneously denying or ignoring the conceptual underpinning of the new thinking: "Defensive Defense" *á la carte*, not as a set menu.

The military/political requirement for this new Europe is stated. It is to find a way of balancing deterrence and reassurance so as to increase military security in the short term as force levels and force densities fall. It may be expressed as three concrete objectives:

1. Improvement of defense efficiency.
2. Enhancement of stability at the four levels of "arms race stability" (i.e., not "competitive strategies"), "political crisis stability" (the Sarajevo syndrome of war commencement, now principally a nuclear problem in Europe), "confrontation stability" (the reduction and removal of temptations for pre-emption), and the "war termination" stability of swiftly achieved "defense dominance" once hostilities commence.
3. Internal harmonization between 1 and 2 and "build-down" arms control.

This chapter next examines three options that currently volunteer for this task: attempting to continue business as usual as if nothing had changed; seeking a high technology formula to compute angles 1 and 2 of the NATO triangle, and adopting military restructuring to enhance defensive capabilities. Each is analyzed in three ways: in its own terms, in terms of the proportion of emotional and formal drive propeling it, and against the three political/military objectives set. Interestingly, only the third is found to conform to the remorseless, paradoxical logic of strategy and to be able to deliver, theoretically, the requisite qualities. It is thought unlikely that it will be adopted in the West for those reasons. Formal nuclear deterrence provides the weakest deterrence and least successfully conforms to the three objectives, but it has the strongest emotional drive. It has outlasted its time for usefulness in European affairs but is expected to continue to be a source of friction for some time.

However, it is observed that probably no military option can offer any assurance of being able to prevail on a European battlefield of the 1990s, for quite different reasons of insurmountable logistical paralysis. This is no cause for concern for two reasons. First, the need now is less of a guaranteed ability to arrive through battle than of an ability to travel hopefully toward a nonmilitary destination. Second, in quality and in quantity the evidence is mounting that dramatic change in the WTO doctrine and deployments make stable, comprehensive disarmament a realistic possibility. Again, the restructuring option offers the best gains. Again, it is thought unlikely that this will sway opposition.

Through the work of Western Sovietologists, a final section examines the progress of military restructuring and reform in the WTO. This suggests that the major doctrinal change of the May 1987 Berlin WTO meetings was indeed major and that it signified effectively the abandonment of a requirement to prepare for the eventuality of not losing a world war. That means giving up the requirement to evict the US from Europe. This is consistent with the unilateral reductions announced in Gorbachev's UN speech of December 7, 1988 and in the Shevardnadze proposals at the opening of CFE. It further appears that since that time, Western ideas of "defensive defense" have become active within the General Staff debate. In this manner, ironically, they are returning with the most power to influence Western strategic assessment. It is predicted that major structural changes in WTO deployments in Europe are to be expected shortly.

The chapter concludes by restating its implicit theme that military restructuring, although essential for the successful achievement of the present transitional military-political phase in international affairs, has no leading role thereafter. The Western official community and its "nonoffensive defense" critics are both guilty of believing too much in military/technical solutions to "Grand Strategic" problems.

BACKGROUND

For more than a decade, an interloper has stalked the corridors of power. "Nonoffensive defense" is most widely understood as a series of originally and still mainly West German ideas about ways to harness new military technologies for two linked purposes. The first is military and political, in that order, and advocates basically blocking or barrier operations at the inner German border, intended to stop a Warsaw Pact *blitzkrieg*. The second is political and military, a reverse order to the first, and is to make the nuclear component of NATO strategy redundant. It is a moot point which purpose came first in order of priority. In both cases, the ideas are highly specific in time, place, and purpose. This remains the case even when, as now, "nonoffensive defense" ideas have gained some prominence in countries of the WTO. "Nonoffensive defense" is proposed for adoption either unilaterally or bilaterally by the confronting alliances. Whereas, as argued below, these ideas can be powerful, they do not represent a fundamental revolution in military affairs and they are generally ill suited for export to other parts of the world. "Nonoffensive defense" is a specific idea in European land warfare. It has naval and air reflexes but not autonomous roles.[1]

"Nonoffensive defense" (also answering to many other names, at least 12 of which have been identified by that most assiduous of detectives, Ambassador Jonathan Dean),[2] has been trying to get rid of its inverted commas, but it has failed. This is because until very recently, insofar as the military authorities of NATO and the WTO ascribed any military meaning to this idea, it was as the antithesis of all good soldierly practice. They took it to mean, by folly or by failure, self-induced immobility: the Maginot Line or a repeat of the disasters of June 22, 1941, respectively. They detested it and certainly had no intention of dignifying the criminal interloper by striking off its shackles. But mainly they viewed "nonoffensive defense" in another light.

The Westerners observed that the idea blossomed in or near the dreaded "peace movements" of early 1980s. They stared with horror at the antinuclear purpose, and they recoiled instinctively. The Easterners knew an opportunity for *maskirovka* (attack by deception) via the "indirect approach" of subversion when they saw one, and therefore were happy enough to back anything that seemed likely to attack NATO's nuclear forces asymmetrically. Of course, it went without saying that nonoffensive defense, or NOD, wasn't to be applied to the WTO, whose offensive defense, perfected during the massive conventional forces buildup in Europe since 1966,[3] was the opposite of immobility and was in any case, by definition, utterly friendly.

Faced with such uniform hostility on the one hand, and such alligator's encouragement on the other, most proponents of ideas from the Land of NOD, who were Western liberals and socialists and in any case not military people, tended to do two predictable things. In the first place, they bickered disputatiously about what the "true gospel" of NOD actually was, and who held it, and in the second place, they overstated, sometimes recklessly, what the adoption of their version of NOD could do. So the inverted commas remained and only served further to discredit the idea in the "official mind." All this was a pity, if not surprising, and all this has since 1987 changed beyond expectation and recognition or, where it hasn't, must do so swiftly.

The military establishments of both East and West now accept the likelihood of the most far-reaching transformations in their military postures and doctrines in Europe since 1945. They expect such transformation soon. Those who are prudent, like the Soviet General Staff and the Bundeswehr, are beginning to ponder deeply, what this will mean for doctrine and deployment. Those who put their faith in Stalin's Return to save them from such travail (notably in the British Army) will be caught unprepared. Like the foolish virgins in the parable, they will be without oil in their lamps if, as is most likely, other circumstances than that Return eventuates. They misstate the choice. It is not between better and worse forms of defense. It is not between keeping up your guard or surrendering. It is between prudence, vision, and imagination on the one hand, and being an ostrich on the other.

Meanwhile, such abrupt change has caught the Western proponents of military restructuring rather by surprise. After knocking for so long on unyielding doors in the corridors of power, when those doors, especially the door to the Soviet General Staff, have suddenly swung open, they have landed unceremoniously, all of a heap, inside. Many are still rather disoriented. Some continue to plug their same messages of the last 10 years, as if nothing had changed: an incongruous duet with those British generals who are the unreconstructed apostles of unconstrained mobility. Some seem overawed by the flush of attention from ministries and generals and too easily compromise their basic principles in exchange for the chimeric fairy gold of "influence" and "access."

In short, there is a need for everyone to get their acts together, and swiftly, for it is an underlying presumption of this chapter that 1990–91 will probably be the critical years for "nonoffensive defense." In the coming year, if the shackles of the inverted commas can be knocked off, the best of the ideas of "nonoffensive defense" may make their contribution to the transformation of Europe, which is now in train. If they cannot, they will not. But the transformation will continue and be the poorer for it. This is because, for the first time in a generation, the challenge of Europe is being engaged simultaneously, at the Grand Strategic level, by the four geopolitical groups with a particular interest (and the three with an autonomous military stake) in it: the Western and Eastern Europeans, the Soviets, and the Americans. The experience of history should teach that such a momentous alignment is wisely recognized and valued for what it is, and exploited without delay.

1990–91 is likely to be the decisive time for NOD because its natural and primary forum has become the CFE disarmament talks in Vienna. If all goes well, these negotiations will have starkly and widely visible, politically tangible, and money-saving outcomes. They will transform military establishments and military-industrial complexes. They will also stabilize the foundation upon which the nuclear disarmament negotiations, those both technically unconnected (START and the resumed test bans) and formally linked (the follow-on European Nuclear Forces negotiation as agreed at the May 1989 NATO summit), may then more solidly stand. For the future of Europe, they are therefore the most important set of East/West negotiations since the beginning of the Cold War. In 1990–91 CFE I will probably conclude and the agenda for CFE II will be shaped, each phase having consequent interaction with the force-planning processes and military doctrines of the two alliances.

The best of NOD has much to contribute to the successful execution of this most vital and most difficult step in the new East/West *pas de deux*. However, it is not at all certain that the Noddist choreographers are capable of meeting the challenge. Among proponents, there is a frustrating and common failure to perceive the *conceptual level of applicability* of their ideas, and in consequence, two common causes of breakdown. One is the "category mistake": inappropriate application of an idea that lives in one level, usually a low level, to another, usually a higher level. The other, a consequence of the first, is "entrapment."

Professor Hylke Tromp describes a cruel paradox. He believes that proponents of NOD pay an unacceptably high price in surrender of intellectual freedom. For by arguing their case technically, as they do, they lose the opportunity to engage Grand Strategic issues in commensurate conceptual scale, and so they render themselves unable to engage the very issue with which they most desire to grapple.[4] Soviet military thinking distinguishes with care between higher order "military-political" and lower order "military-technical" levels. The flow of command is downward. A constant failing in Western strategic thinking has been confusion of these two levels and especially allowing technology to drive strategic rationale. Proponents of NOD have also been guilty of the same failing. Most of the earliest ideas of "nonoffensive defense" were formed and exist at the "military-technical" level, yet they are offered as "military-political," even strategic solutions, where they are swept away by Tromp's remorseless broom. Yet, it will be shown, new ideas about "defensivity" *can* be legitimately active at the Grand Strategic level, but not directly. Therefore, to avoid Tromp's Broom, the primary task in assessing the potentials of NOD is more clearly to understand the dimensions of the Grand Strategic environment in which they live now and will act.

A NEW VISION OF EUROPE'S FUTURE

After the Battle of Waterloo in 1815, it was generally agreed among the victors that Napoleon was a mistake that must not be permitted to happen again. An efficient, nonmilitary way of regulating the affairs of the Great Powers of Europe had to be devised. The Concert of Europe worked for two generations (during which the British, and subsequently the French, were extremely busy carving out colonial empires, first by informal and later, during the so-called Scramble for Africa, by more formal, military means). The Concert collapsed. New Franco-German enmity led to the war of 1871. A further, rickety attempt at nonmilitary adjudication was attempted. Meanwhile, insufficiently studied in Europe, the first war of the twentieth century had been fought in the United States in the 1860s. In the systematic application of industry to warfare, in the adoption of Sherman's principle of limitlessness in war, overriding the older "just war" tradition of *jus in bello*, the American Civil War prefigured the concentrated violence that was preparing itself for the new century in the foundries and workshops of the industrializing Old World.

The Anglo-German naval arms race—the first industrialized arms race—helped to inflame war fevers. But when it came, the first "hot war" phase of the great European "civil war," 1914–41, burst upon the heads of those who willed it with unanticipated destructiveness. Millions of lives and four empires (German, Austro-Hungarian, Russian, and Ottoman) were forfeit. One (and in 1918, in Germany, nearly another) imploded in Bolshevik revolution. Europe was ripped apart, politically and demographically, in the war and in the peace. At Versailles, as J. M. Keynes presciently observed in *The Economic Consequences of the Peace*, Europe gained nothing in the long run by cutting off its German nose to spite its continental face. The double injury festered during the Cold War of the 1920s and 1930s and oozed fascism. Keynes' book sold well in German translation. Another empire (the British) sustained grave internal damage at the hour of its greatest apparent success: overextended in war, pressured by the Depression. It was a graphic illustration of the paradoxical basic guideline for those who possess and wish to keep their empires: "If you use it, you lose it."

The Japanese, the second-rank victors of World War I, unexpectedly freed by it of their Meiji debts to the industrial powers, were also freed to contemplate the virtues of possessing an imperial cushion when the bad times next returned, with the depression. In the staff college teachings of Ishiwara Kanji in the late 1920s and, after the Mukden Incident, in the escapades of the Kwantung army in Manchuria during the 1930s, they rehearsed. And the United States, sole victor of the first rank in the war, became the reluctant Osymandias who, in refusing his crown, helped to plunge itself and the world into depression. It took the shock of witnessing the second "hot war" phase of the European civil war, compounded by the experience of Japanese attack as they sought fruitlessly to achieve by war that Co-Prosperity Zone they were to develop so successfully by non military means, later, to cause the US to emerge from isolationism. The World War II rekindled the fires under the cold boilers of America's depressed economy and generated a head of steam that rocketed America out of the Depression and into the heady (if temporary) economic and political abundance of the Long Boom.[5]

After each of the great destructive catastrophes, just as after the Napoleonic Wars, the international community sought, with renewed vigor, to devise a new means of settling interstate disputes without war. After World

War I it was the League of Nations; after World War II, the United Nations. Each war was bigger; each attempt, more strenuous; And each attempt failed. The UN was conceived in knowledge of the reasons for the failure of the League and was born under the shadow of the atomic bomb. In 1945 a fierce hope that this time a phoenix of peace might rise from scorched ashes more extensive and terrible than ever before dictated the pragmatism that gave the veto to the Security Council Permanent Members (and nominated those members) and equal membership of a General Assembly to all sovereign, independent states. Resolution I(1) of the first General Assembly in London in 1946 expressed the idealism that, fortified by common purpose, the new atomic dimension in international affairs might be, from the outset, controlled in a new way. It set up the Atomic Energy Commission. But a combination of Baruch's punitive redrafting of the Acheson-Lilienthal proposal, the young Gromyko's faithfully wooden-mouthed expression of Stalin's reciprocal hostility, and the first Bikini test blew away that chance. The world settled back into the familiar, confrontational mode, and the UN became for many years just another shuttlecock in that game.

Four strong pillars held up the sky of the postwar world. All were made in America, although some were later further embellished in Europe. The first was a war prevention system called "nuclear deterrence," created, *ex post facto* as a consort for the Bomb. The second was a global alliance for the containment of Communist expansionism, whose charter texts were given in George Kennan's 1946 Long Telegram and in Paul Nitze's 1950 NSC-68 and whose main expression was NATO. The third was a willingness to intervene by force in the nonwhite and/or colonial (later "developing," later "underdeveloped," later, "Third") world to prevent leftist revolutionary change, whose first failure was the defeat of the Kuo Ming Tang by the Communists in the Chinese civil war in 1949 and whose revenge was in Korea. The last pillar was a dollar-based international market economy. Since the end of World War II, America has successfully defied economic gravity by three devices successively, to maintain its military and political expenditures without commensurate domestic taxation. All exploited the dollar's continuing postwar privilege in an expanding world market of which the U.S. economy represented a decreasing proportion. In order, the three balls in the juggling act were the Bretton Woods system of 1944, followed, after August 15, 1971, by the Nixon floating exchange rate, and then, under the president who sought a balanced budget amendment to the Constitution, by the biggest deficit that the world has ever seen. Reagan's deficit was, in the judgment of the most accurate of the predictors of the October 19, 1987, crash, "the most irresponsible fiscal behavior in [American] history."[6]

For the Europeans west of Churchill's Iron Curtain, NATO was home and NATO defined their hope and their salvation. In the crisp words of Lord Ismay, the first secretary general, NATO's role was "to keep the Russians out, the Americans in, and the Germans down." For Europeans to the east, after what Hungarians poignantly call "the year of the turn" (1948), life was quiet despair and gray malevolence under the imposed regimes across Stalin's *glacis*, punctuated with an occasional, wild efflorescence of hope in Budapest in 1956, Prague in 1968, Gdansk in 1980; each crushed. Only upon his Prussians could Stalin rely. A "sandwich" deployment was used in the Warsaw Pact's order of battle: Soviet or East German units flanked those of other "allies." The British in India did the same thing on the Northwest Frontier with native levies. Ever since the Mutiny, any seasoned Indian Army officer could have told you that that was what you did with unreliable sepoys, so that you could stick your bayonet into the enemy or into them, *pour encourager les autres*, as circumstance demanded. The Cold War world seemed hard and frozen solid. It was noticed, with some private disquiet on both sides, that the most efficient exponents of capitalism in the West and the most faithful and efficient Marxist-Leninists in the East were both German. No stronger argument for the perpetual division of the Germans could be adduced.

This reminder of history gives necessary perspective to the opportunities we now confront. Within the space of a decade, erosion of the four pillars of the Cold War world has become, in all cases, visible fracture. Stalin has gone, giving us in his place Gorbachev.[7] From 1968 to 1979, the Americans wallowed in the pain of Vietnam and Watergate and the Iranian hostage crisis until the genial Reagan offered them a cheerful holiday from reality (kindly paid for by the Japanese), from which they are now returning, hopefully refreshed, to find the world much changed, and nowhere more so than in Europe.[8]

At its beginning, the European Economic Community looked likely to become an embellishment of the second pillar. The haste with which West Germany was rushed into CECA (La Communauté Européenne de Charbon et Acier—The Coal and Steel Community, which was the trial run for the Treaty of Rome) seemed all of a piece with the American military governor, General Lucius Clay, forcing the new West German currency in 1948 (which led directly to the Berlin Blockade) and the refusal of Marshall Aid east of the Iron Curtain: all were means to crystallize the division of Germany as swiftly as possible. From the French perspective, the division of Germany was also expected to deliver, and the EEC to seal, a permanent political predominance for Paris. The objective was the same as Clemenceau's when he sought the "pastoralization" of Germany at Versailles, only the means were more sophisticated. Yet the EEC hasn't turned out to be like that at all.

Compelled by their conquerors to abjure "distractive" investment in matters military, the Germans, like the Japanese, focused their prime asset, their remarkably skilled and hard-working people, upon "contributive" economic activities.[9] Also, for understandable reasons of collective psychology, they preferred to forget the recent past, climbed into their Volkswagens, and drove off into the new dawn of the German "economic miracle."

Throughout those years of forced, formal political impotence, during which the FRG was becoming the economic center of the new, emerging Europe, two threads ran through German foreign policy, one eventually visible, the other hidden. The visible thread was *Ostpolitik*: Willi Brandt's conviction that the best way to neutralize the threat from the East was not by containment (pillar two) but by constructive entanglement in a cat's cradle of economic involvement. *Ostpolitik* was always regarded coolly by the European supporters of pillar one, in Paris and London, and with hostility in Washington. It was tied to the hidden thread, which was always there and has in the last year become boldly visible: reunification. Both were spun from the same wool: a belief that military power is much overrated as a source of usable influence, compared to economic strength.

The continental Europeans responded to the oil price shocks of the early 1970s quite differently from the Americans. On the one hand, given the faster replacement cycle of capital goods and higher rates of investment in infrastructure and in design, European products became quickly more energy efficient than American. They also had railway trains. On the other, Western Europe looked around for a more politically reliable source of hydrocarbons than the sheiks. By 1976 the USSR was Western Europe's largest single supplier of oil and natural gas. The income from those sales bailed out Brezhnev's policy of *laissez faire* corruption. It also built the "turnkey" fertilizer factories that gave Gorbachev, the first secretary of Stavropol in the Caucasus, the wherewithal to preside over bumper harvests when artificial fertilizer was first applied to the black soils, a lucky stroke that drew him public attention and promotion to Moscow on Kulakov's death.[10]

The de facto advance of *Ostpolitik* drove even Margaret Thatcher, against her instincts, into conflict with Reagan when he tried to embargo equipment for the Siberian Gas Pipeline project in 1981. The Europeans would not countenance it and she could not defy them. It was the first of many such defeats within the EC and within NATO. *Ostpolitik* has now created a situation where more than 50 percent of the DDR's imports come from the FRG and 60 percent of its foreign trade is with the FRG, raising the question of what the real criterion of reunification really is. It has produced a moment when Hungary is suing for a backdoor association with the EC, riding upon the back of the special relationship with Austria, in Austria's negotiation. ("We shall remake the Austro-Hungarian Empire," joked a Hungarian acquaintance, "but this time, we shall get the capital in the right place!") It had produced the possibility by the summer of 1989 for Gorbachev to plead before the Council of Europe at Strasbourg, with unembroidered urgency, for the EC members to realize that the success of *perestroika* depends to a significant degree upon their willingness to inject the steroids that will develop rapidly his economic muscles, an argument that the first non-Communist government of the generation, led by Solidarnosc in Poland, has echoed with some desperation; and it has produced the fact that the USSR in autumn 1989 held more Forex credits from the FRG than the Soviet economy can manage to take up.

In short, since 1972 by force of circumstance, by force of argument, and by force of growing economic preponderance, the German perspective on the power and utility of the *ostpolitik* route to security has prevailed in Europe. While recognizing German economic *force majeure*, the French have cleverly built in their position—physically. They have practised good old nineteenth-century railway imperialism. As we approach the next age of the train, the hub of the European high-speed rail network for the turn of the millennium will be Paris, from which the TGV (Train Grande Vitesse) lines radiate across the continent and shortly will pass under the English Channel into the depressed northern island provinces. The French have, at long last, reversed the result of the Battle of Waterloo. It is a fact already fixed upon the map!

They have also been happy to support the German drive for more trade with the East for another reason. In 1992 comes the completion of the Single European Market, at which time goods lose any national protection in their domestic markets. The French also wish to get on with business to the East to help buttress the more fragile sectors of their domestic manufacturing industry against the buffeting to be expected after 1992, notably from the Germans and the "Japobrits," by securing a sheltered export position. So Prime Minister Rocard's public call in late 1988 for pan-European participation in the making of the new decisions toward the East serves only to underline the convergence of Franco-German interest. Genscher and Mitterrand make together a new *ostpolitik*, with Italy as a junior partner, and agree to share the profits, cutting out the other Europeans (especially the British), the Americans, and, of course, their main rising enemy to the East—the Japanese. They, nontheless, are creeping into Europe as "Japobrits" via "indigenized" factories set up in low skill, low wage Britian: an ironic reversal of the role that Imperial Britain gave to Japan for 20 years as its China Sea deputy in the 1902 Anglo-Japanese Naval Treaty.

These considerations bear directly upon the assessment of "nonoffensive defense" in three, convergent ways. The first is that "1992" has acquired inverted commas also, for it is not only a date, it is also an event. The European Economic Community is well on the way to becoming the European Community. Article I of the Single European Act states an objective ". . . to contribute together to making concrete progress toward European Unity." The hopes of political union, nourished at the outset by Churchill as much as by Monnet, languished while De Gaulle soothed French injury from defeat and guilt over collaboration with strong draughts of *La Gloire* and while Germany was still weak, but have gained sudden momentum in the last five years. Certainly, there is an inherent contradiction between closer union among the founding members and enlargment of the Community, now even contemplated in some form across the Iron Curtain. Equally, there is tension between the approach of a "Europe of states" and a "Europe of regions"; and there is some dissent from the core values of European social democracy. The project is afoot in ways that were surprising only a few years ago: *L'idée Européenne* is politically malleable, potentially volatile, and autonomous for the first time in a generation.

Second, this brief history shows not only *that* but *why* German ideas have come to dominate this European debate about Europe's future at this sensitive time. Wellington, the Iron Duke, understood precisely how to play London's role as the third corner of an Anglo-Franco-German triangle of power. But for good or ill, the British corner has thrown in the towel. The Boot would give his pygmy successors, one more opinionated than most, especially, the boot for their narrow-minded incompetence! A Franco-German axis will now principally determine the next stage in European development. Germany's return to formal power has been an object lesson in a central feature of the times: the sharply diminishing utility of the old means to power—possession of abundant armed force—in contrast to the rapidly increasing salience of the new—knowledge and the ability to apply it to productive ends—in a world involuntarily suffused with information by the electronic revolution in information technology.[11]

Third, the simple fact that most of the seminal thinking about NOD has been German should now, of itself, cause the reader to stop and think a little harder about what it may mean to Europe that Germans are so interested in it.

The new vision of Europe's future may be succinctly stated. It has three aspects. The first is the most astonishing. The European civil war has, throughout the twentieth century, been its most consistent source of chaos. Each time that there has been a fresh attempt to construct a system of collective security, searing memory of the misery and mayhem of the previous "hot" phase in the civil war has been the spur to action and quarrels over Europe, or emanating from Europe, have been cause of subsequent breakdown. Now, for the first time, we have an opportunity to build a system of collective security *that has not required the spur of war*. The importance of this fact cannot be overestimated. The chance comes unpredicted and unexpected. Reasons for this have been offered above. They define the second aspect. They reshape axiomatic assumptions about the nature of power.

The challenge to NOD, more closely defined, resides in this second aspect. Franco-German hostility led three times to war since 1870. Yet now no one calculates the military balance between France and Germany. The present task is to find a way to achieve just such a permanent improvement in the general European military-political climate by devising a European system of collective security as a feature of a general global system.[12] The CFE negotiations offer a way in. As we enter by that door, the third aspect flashes a warning. Military forces are still deployed at historically unusually high densities and levels of readiness. While hoping for the best, one must always be prepared for the worst. But that preparation must not, as it has too often in the past, stifle moves to escape from the self-fulfilling circle of threat. Finding a way to balance deterrence and reassurance is the "military political" and the increase of military security in the short term as force levels and force densities fall is the "military-technical" theme that must be choreographed for the new European *pas de deux*. NATO's Comprehensive Concept states as much.[13] But as the tempo increased during 1988–89, it has become increasingly clear that the alliance's thinking cannot escape from the familiar circle of threat. Its own stock of ideas is not up to the task of masterminding the breakout. This is where NOD can help.

HIGH NOON FOR "NONOFFENSIVE DEFENSE" IN THE WEST

The Vienna talks on Conventional Forces in Europe have grown rapidly in visibility and it has been suggested that the CFE negotiation is potentially of even greater significance than is widely appreciated. CFE is the successor to a string of cognate negotiations, notably the Mutual and Balanced Force Reduction negotiations, which dragged on for years.[14] It has become commonplace to compliment CFE by pejorative comparisons with MBFR. That is a mistake. MBFR was very successful: nothing happened for 15 years. That was certainly what the USSR intended to be its outcome at a time when Soviet doctrine called for a prolonged conventional forces buildup in Europe. In contrast, the speed of movement in the CFE has been startling. In the first three months of the talks, issues of principle have been resolved that, in the opinion of the former U.S. ambassador to MBFR, place the parties about where they were in the INF negotiation after four years.[15] The main reason for this, of course, has been the spectacular change in the WTO position from the dog days of MBFR (more in the final section). Suffice it here to recall that the unilateral cuts announced in Gorbachev's December 7, 1988 speech to the UN equaled those proposed for bilateral negotiation in the Budapest Appeal of 1986.[16] The point is simply that in a profoundly political transaction that is conducted in military currency, a change in strategic objective may facilitate huge changes in detail at the technical level.

There is deep ambivalence in the response of the Western military community to this train of events. One voice seems to be unequivocally warm and welcoming. NATO's Comprehensive Concept claims expansively that "In particular, the concepts of stability, reasonable sufficiency, asymmetrical reductions, concentration on the most offensive equipment, rigorous vertification, transparency, a single zone from the Atlantic to the Urals, and the balanced and comprehensive nature of the CSCE process, are Western-inspired."[17] Although exaggerated, this claim is not entirely without cause. A consistent feature of WTO diplomacy on conventional forces has been to recycle Western ideas as the vehicle for its own proposals. This leaves moot the question of what is *pokuzuka*—show without substance—and what is not, addressed below. Further enthusiasm was evinced when, for the May 1989 Brussels Summit, the Bush administration decided to "go pro-active." It overcame the initial reluctance of other parties, notably the British, and proposed the inclusion of aircraft, helicopters, and manpower as variables in the calculation. Perhaps encouraged by the beneficial effect that this prudent concession had upon public opinion, the principles of the NATO proposals on limits for the categories announced in May, promised for September, were in fact published on July 13, 1989, before the summer holiday. (NATO's proposed limits on land systems—tanks, armoured troop carriers, and artillery—had been published on March 6).

The July document went even further than being cooperative. It suggested that once these reductions had been agreed, presumably in the first phase of CFE," . . . we would be willing to contemplate further steps to enhance stability and security in Europe, such as . . . the restructuring of armed forces to enhance defensive capabilities and further to reduce offensive capabilities."[18] What could be clearer than that? It seems as if the principles of NOD have been accepted, that the interloper has had the shackles struck of and is seated in the office.

Early the next week, the director general of training and doctrine (army), Major General Pollard marched to a different drum. For a distinguished gathering at a British Army conference, the quotation he chose as the motto for his theme paper on the future defense of the central region was von Manstein's "Passive defence . . . a policy that usually leads to the defeat of the weaker party."[19] Pollard's is a vigorous and subtle essay. Furthermore, it is a considered statement on the record from the general with express responsibility for strategic thinking in the British Army, asked to look at the debate about "defensivity."[20] He analyzes what he sees to be the spectrum of concepts of operation available for the European theater. They range from passive/defensive defense/static/positional across to active/offensive defense/maneuver/deep penetration (the synonyms are his).[21] For him, both the trench warfare of World War I and the Maginot Line of World War II stand to the passive end of the spectrum. The German (and, although not mentioned, presumably the Soviet) *blitzkrieg* stand approved of as sound military thinking, at the active end. Pollard's point is made most trenchantly by his contrast between the mistake of the Israelis' passive defense at the Bar Lev and Suez lines, retrieved by the brilliance of their maneuver warfare in the Yom Kippur War of 1972, which illuminates how ". . . maneuver tactics or maneuver elements have nearly always been prerequisite for victory or success in defending one's territory."[22] At the same conference, the late CinC of the British Army of the Rhine gave another powerful paper that complemented Pollard's. General Sir Martin Farndale's exposition of what he called the highly personal art of command pivoted upon the commander's contemplation of the battlefield with a view to identifying his time and place of opportunity to switch to the offensive and win. Quoting *The Armoured Division in Battle*, Vol I, Pt II, Pollard consolidates the point that ". . . offensive action is fundamental to this phase of war."[23]

But the nub of Pollard's objection to what he calls "passive defense" is at the political, not at the tactical level. He believes that adoption of passive/positional/"defensive" defense, such as the Maginot Line, encourages complacency in the populace. However, of greatest concern, ". . . it is the philosophy of defensive defense that is most dangerous. Defensive defense denies the conventional offensive force elements and thus (sic) questions the validity of deterrence . . . the passive defense option does not provide adequate deterrence and a more robust operational concept is required."[24]

This statement repays a close reading. It contains four contentious assumptions that are stated as fact. One is empirical, one conceptual, and two are logical. The empirical assumption is in the characterization of "defensive defense" as a latter-day Maginot Line, lacking mobility. This is an error corrected below. The conceptual assumption is of the indispensability of the offensive to "finish" the war (and it is moot what "finish" means, as the history of the European civil war earlier this century reminds us). The first logical assumption is smuggled within the sentence and is of a special, positive correlation between conventional offensive force and the "validity of deterrence" (which implies a converse and inverse relationship between "the validity of deterrence" and other, unspecified forms of force). It is unclear what "validity" means, but suggests, a circumscribed definition of deterrence. The other logical assumption is of a stated negative relationship between passive defense and "adequacy" of deterrence (which is presumably different from validity), again implying a converse and this time positive relationship between active defense and adequacy.

Pollard's formulation appears to be oblivious to the very different qualities of deterrence by punishment and dissuasion by denial, and to the very considerable inherent problems with the former (loss of initiative to the enemy, basic credibility etc.)[25] Nor, equally surprisingly, does it appear to appreciate how the paradoxical logic of strategy works. When the power of weapons and formations transcend what Edward Luttwak calls the "culminating point of success," *more* becomes *less*. That this is not understood is because of the illegitimate if understandable use of a linear logic drawn from common sense (Big Bomb equals Big Deterrent; Bigger Bomb equals Bigger Deterrent). Common sense is a dangerous guide in the realm of war. The essential logic of war is paradoxical.

The remorseless and paradoxical impact of overreaching the culminating point of success is deeply offensive and perplexing to the conventional military mind because it does not move, it removes the goalpost of "winning," and it questions the working rule of thumb that you win by maximum destructive effort. But the paradox is to be seen at every level. At the lowest scale—the technical—too excellent a weapon, concentrating within itself too much military efficiency, the more swiftly summons up effective countermeasures to blunt it. At the operational level, overreaching campaigns help produce the *opposite* of what is intended: The German High Seas Fleet produces British engagement on the continent in a triple alliance *with* Russia and France and *against* Germany—the reverse of all reasonable expectation, including in Britain, before 1902. The RAF area bombing intended to destroy German civilian morale instead stimulates German war production and increases the flow of labor into it. At the largest scale, that of Grand Strategy, nothing fails more reliably than decisive military victory. Thus Operation Barbarossa starts the road that led back to Berlin. Thus Pearl Harbor seals the doom of Japan's military expansion. Thus the Six-Day War starts the road that has led to the Palestinian *Intifada* two decades later. There is no reason to think that success in the tremendous, primarily nonnuclear *blitzkrieg* across Western Europe, which from 1966 to 1987 was the minimum Soviet military objective for the first phase of a world war, would escape the claws of this logic and mounting evidence that appreciation of this was part of the Soviet decision to change their strategic objective.

This renovation of basic strategic thinking is an essential part of striking the shackles from NOD and from NATO alike, but it is still beside the point in terms of present realities in NATO, as evinced in General Pollard's paper. What is to the point is that his logic and assumptions about deterrence closely parallel the "conventional wisdom," especially in the nuclear field. There is, in fact, a further level of meaning to be retrieved from Pollard's text. He sees that not only conventional offensive but by extension tactical nuclear operations find no role in alternative defense proposals. As mentioned at the outset, an antinuclear motive has lain within such thinking from the start especially in West Germany.[26] A desire by those who believe in them to protect nuclear roles in Europe is an important reason for the offensive defense against alternative strategies, although among those who understand the meaning of the word *diplomatic*, it is felt to be politic not to say so too openly. Such belief is held very strongly.

Therefore two interim, practical conclusions emerge from examination of the Western ambivalence about alternative defense within the context of the

CFE. The first, which is a military-technical fact although not a military-technical judgment, is that one man's offense is another man's defense and that it is hard to alter that perception at the technical level. The second is that CFE is potentially vulnerable on two fronts. It could be shipwrecked not only by an East/West breakdown, but also by West/West disagreements, notably on nuclear issues. Given the critical role that CFE has in the larger game of European construction at present, such a breakdown is to be avoided at all costs.

Nuclear issues operate at a practical but more powerfully at an emotional and symbolic level. The friction generated within NATO over SNF modernization confirms this. It was, in Paul Warnke's cryptic description, the wrong battle on the wrong issue at the wrong time. Why then was it fought? The battle was fought largely because Thatcher wished to fight it. She wished to fight it not because any concrete decisions on replacement of the Lance missile were really necessary, but because she sensed the threat to the "validity of deterrence" (understood as her view of deterrence), which was posed by the growing consensus that the course of events was rendering existing, let alone new nuclear weapons in Europe, and especially in the FRG, less necessary. She lost every major position she tried to defend, overwhelmed by German *force majeure* at each point. The conflict was an echo of the larger debate between the four pillars and *Ostpolitik*. What differed was that it was not, in the final analysis, an issue for which the Americans were willing to climb into the trenches and jeopardize their relationship with their main European ally, the Germans; and so without a new Pershing (human or ballistic) to reinforce her, Thatcher was forced to retreat. Because of this bruising episode, and its continuing potential to disrupt, a further word about tactical nuclear strategy will be required when we assess the ability of NOD or any of the other military options available to deliver "adequate deterrence" and other necessary things, in a moment. The stakes are high, and if NOD fails to make its contribution at this time, it is hard to see how or where it will have an equally important opportunity. So this is high noon.

The immediate problem for NATO in CFE is thus seen from NATO's official perspective to have three corners.

1. How to manage cuts that are probably irresistible and are now likely to be considerably more than the cosmetic adjustments that were originally expected. "Manage" means both to think about and to act upon. "Irresistible" means that they are driven by more than arms control.
2. How to make such cuts without damaging the present formal NATO doctrine. "Damage" means a public admission or undeniable public exposure of inadequacy. In other words, how to confront a new situation without any fundamental change of perspective: without new thinking.

3. How, at the same time, given that the present formal doctrine may not do (a) to draw benefit from new thinking that may help to do so, while always preserving objective, and (b) this means simultaneously denying or ignoring the conceptual underpinning of the new thinking: NOD á la carte.

NATO's "noncomputing" triangle presents advocates of NOD with the ticklish dilemma of whether or not to accept this description of the issue. If they do, they may win smiles but risk being lost in the triangle. They risk additional criticism from the Western establishment but may remain afloat if they don't.

Yet there is some converging consensus about objectives as between the official community and the Noddists, although little about the means to achieve them. The consensus lies within the third aspect of the new vision of Europe—finding a way to balance deterrence and reassurance so as to increase military security in the short term as force levels and force densities fall. It may be expressed as three concrete military/political requirements:

1. Improvement of defense efficiency.
2. Enhancement of stability at the four levels of "arms race stability" (i.e., not "competitive strategies"), "political crisis stability" (the Sarajevo syndrome of war commencement, now principally a nuclear problem in Europe), "confrontation stability" (the reduction and removal of temptations for pre-emption), and the "war termination" stability of swiftly achieved "defence dominance" once hostilities commence.
3. Internal harmonization between 1 and 2 and "build-down" arms control.

This is an issue that resides within the difficult and hitherto unequal relationship between force planning and arms control in the West. The definition of what composes the operative terms "defense efficiency" and "defense dominance" in the prescription is fundamentally in dispute.

WEIGHING THE OPTIONS

There are three ways forward for the West at this juncture. Each has its own logic, each may be criticized in its own terms, and each must be criticized according to the larger criteria set by the new vision of Europe. Furthermore, each option may be advanced in formal and in emotional terms. The degree of each in each case must be gauged.

"Full Speed Ahead and Damn the Torpedoes!"

Confronted with obstacles, to soldiers and politicians, there is a certain flamboyant attractiveness in decisive action that gets the matter resolved one way or another. Admiral Farragut's famous order in the Battle of Mobile Bay (Alabama) in 1864 even has its counterpart in the modern diplomatic history of Europe. When it was reported that strikes were threatened by the Left and that public opinion polls were registering massive majorities opposed to his decision to cause the FRG to join NATO, Konrad Adenauer is reported to have replied, "then change the figures!" A recent opinion poll in the FRG registered only 9 percent actively supportive of the NATO strategy of Flexible Response, including the possibility of a NATO first use of nuclear weapons.[27] There are certainly those in Europe but not in Germany, and some in Germany, who favor a similar response to Adenauer's at this time.

For those who hold this view, the logic of forging ahead regardless is compelling. It is buttressed by two memories and a fear. The memories are of Adenauer, and more recently, of the successful emplacement, in the teeth of public hostility, of the Pershing II and Cruise missiles. In both cases, it is thought, subsequent events vindicated the decision: a West German as secretary-general of NATO, and by standing firm, the peace movements were vanquished and the Russians, suitably impressed, stopped playing meddlesome games and conceded the Western demand. The INF Treaty was the result. Even if you can't explain exactly how it worked, you don't throw up a winning strategy. The fear is that any other course of action than Farragut's now would appear as weakness, which would sow the dragons' teeth of more peace movements and would encourage the Soviets to more meddling by *maskirovka*.

This interpretation of cause and effect in recent East/West relations is hotly contended. It is argued by others that in fact it was much more the arrival of Gorbachev and related changes within the WTO that led to the INF accord; by others again, that no monocausal explanation is really satisfactory. Such alternative explanations tend to be impatiently waved aside, which shows that this position is maintained more on an emotional than a formal basis. It is akin to Thatcher's conduct during the SNF modernization issue, to which it is formally kin, too. In similar vein is the bland assertion of the continuing good health of Flexible Response to be found in the Comprehensive Concept. In the last few years, collateral confirmation comes from the defensive tone of an increased number of longer and better informed "essays" in the annual British Statement on the Defense Estimates, on topics such as the nature of deterrence and the exclusive interpretation of recent history given above.

However, it is conceded that successful as it has been, and uniquely able to deliver adequate deterrence as it is, the present strategy faces severe problems. Two in particular preoccupy the military community: the birth rate and the scissors crisis of general defense cost inflation as against stagnant or shrinking real resources. General Pollard's paper discusses both. Demographic predictions show a trough in the number of 15–19-year-olds in the 1990s, which suggests that to attract and to retain well-trained volunteers in substitution, manpower costs as well as the more familiar equipment costs are likely to increase beyond the capacity of the tax base to sustain.[28] Whether or not it is a cause of comfort to beleaguered generals in the West, the same factors are pressing upon the WTO with even greater immediacy. The existence of these pressures has caused Admiral Sir James Eberle to coin the memorable acronym MUD—Mutual Unilateral Disintegration—to describe their likely effect upon the military establishments of both sides, regardless of any reductions produced as a result of disarmament negotiation. If you are going to slip on the MUD anyway, there is virtue in being light rather than overweight when you do so. So these factors compose a compelling motive for engaging in cuts if you can.

The Fallacy of the Last Move—Again

Preoccupation with the threat of Eberle's MUD-slide combines with a desire to compute angles 1 and 2 of NATO's triangular problem to promote a second option.

A vigorous restructuring of NATO's posture in the central region, strengthening interdiction, and especially deep interdiction capabilities is thought to be attractive for many reasons. As Air Chief Marshal Sir Anthony Skingsley has rightly observed, interdiction is a classical and elegant military gambit to which NATO has historically paid close attention, forced to do so by the numerical imbalances against it. The mission of interdiction requires speed, range, stealth, and cunning. It is an intensely active operation as the detection-decision-attack cycle repeats at high tempo. Its objective is to seize and to retain the initiative. These qualities harmonize philosophically with the gospel of mobility. They also express well the dominant ethos of the U.S. Army's modus operandi. Indeed, it was that army's development of the doctrine of Air/Land Battle in the early 1980s[29] that led, under SACEUR Bernard Rogers, to the promotion and adoption as an ACE-wide policy a renewed emphasis upon the cost-effectiveness of interdiction for NATO in the form of Follow On Forces Attack (FOFA).[30]

The possibility of a further, major substitution of a technical for a human means of covering space with force is now made possible by the stunning C3I capabilities already to hand or near to hand (Joint STARS airborne radar, KH-11 and other satellite remote sensing systems) in combination with highly capable, deep-penetrating aircraft deploying new varieties of "smart" munition, and with accurate long-range conventional munitions (ATACMS, etc).[31] The attraction of advanced FOFA to its advocates is that they believe that it could do considerably more than help to shatter a *blitzkrieg*. The following arguments are to be encountered:

1. It might help to relieve the difficulties that the birth rate drop will create by substituting airborne combat power for land forces, shifting the air/land proportions in an integrated strategy.
2. This would relieve some of the pressures for exercising near the IGB and thus, by making NATO less visible, might sooth public irritations in the FRG.
3. In an environment conditioned by CFE and the expectation of cuts, a well-developed FOFA capability would be the best insurance against future Soviet "breakout" as numbers and force densities fall, should Stalin return. Only with a mature FOFA system fielded would it be militarily prudent for NATO to engage in more than cosmetic cuts in classical land forces for purposes of arms control, for only by substitution could the integrity of NATO deterrence doctrine be sustained.
4. Joint STARS can look along the IGB as well as across it. The smart, multisensor, instant and delayed action, antiarmor, antipersonnel, anti materiel, and area denial mines, which are laid by FOFA delivery systems, can create choke points at the Forward Edge of Battle (FEBA) as well as deep behind it. The whole package is thus a valuable enhancement of NATO's capability to fight positional or defensive actions. In fact, without Advanced FOFA, NATO would be ill equipped for transition to an even more visibly defensive posture.
5. FOFA raises the nuclear threshold in two ways. The first is by increasing NATO's ability to fight on the conventional "rung" of Flexible Response.[32] The second is that a pronounced deep attack capability would, by marking a clear threshold in space and in the levels of military preparation, increase stability. NATO could then afford to alter its declaratory nuclear doctrine to a conditional First Use in the event that other than in-place forces move. FOFA would thus improve "political crisis", "confrontation" and "war termination" stability.[33] This should be attractive to sensitised continental European public opinion.

Advocates conclude that there is not only military sense, but also realism in such a reform. These five arguments conform to the demographic, budgetary, and technological futures that NATO faces. Advanced FOFA is also politically wise, for by its military flexibility, it offers that essential political insurance without which it would be folly to pursue the opportunities to obtain deep and asymmetric cuts in WTO land forces presently available. Thus it is the most prudent next evolution in NATO doctrine.

Often in the past, technical wizardry has appeared to offer an escape from a seemingly hopeless impasse. Faith in the ability of meek, bespectacled Clark Kent to leap into a telephone box, to emerge as Superman, and to save the day is deeply rooted in well-schooled comic readers everywhere. The distressing lesson that each resort to a technological fix in the modern East/West confrontation has only served to rachet up the costs, the physical precariousness, and the mutual mistrust of the confrontation appears to be impressively forgotten. This inability to understand the Fallacy of the Last Move is noteworthy. It speaks to the depth of emotional engagement that the issue of Western and especially American technical prowess stirs. There is no obvious reason why Advanced FOFA should be different from MIRV, or Cruise, or Stealth technology, or the many other precedent Last Moves. However, Sharfman argues against precedent that the Joint STARS technology is so far beyond what the WTO could achieve that this Last Move is really checkmate.

Sharfman and Bracken are also both aware of criticism of the vulnerability of the valuable Joint STARS aircraft and of the fragility of so ramified and articulated a data loop. (Sharfman lists nine separate but essential functions that must be provided and that must work well for FOFA to work at all.) Both seek to answer the critics. Sharfman attempts to rebut the charge made in Congress when funding of Joint STARS was under discussion, that it could never perform its mission. The criticism was in the form of a Catch-22 proposition: if it flies high to get radar coverage, it will be shot down; if it flies stealthily to be survivable, it can't give the required coverage. Sharfman thinks that the issue is not as stark as this and that partial performance by the fleet of 22 aircraft now authorized can meet a satisfactory operational requirement.[34]

Bracken offes a most ingenious attempt to proof the C3I architecture against breakdown. However, his proposal also fails to convince. Both a horizontal expansion of intelligence assets and a raising of the executive decision point up the command tree fall victim to the paradoxical logic of strategy: the former because nodal points for high quality data processing cannot be avoided; the latter because they are them.[35]

Bracken is correct that if the enemy is confident that Advanced FOFA is both invulnerable to pre-emption and will work efficiently in battle, it would pose a credible threat, which, being independent of classical conventional force numbers, would remain stable under crisis at reduced force density and numbers. But unfortunately there is no good reason to believe either of the conditional statements in the previous sentence and every reason to believe that such a restructuring would reliably decrease reassurance by stimulating an arms race and would raise the bonus for pre-emption. The central, inescapable technical problem with Advanced FOFA is that it conclusively surpasses the culminating point of efficiency.

Judging Against External Criteria

Both of the above options have been assessed only in their own terms. One flimsy case can be made for new technology staving off the MUD-slide and another for Advanced FOFA maintaining crisis stability under a regime of numerical reduction at the cost of increasing other instabilities. However, they must also be measured against the higher order strategic requirements of the new Europe. Because the one option is an extension of the other, they may be taken together. Both have a high proportion of emotion in their construction and are accordingly stiff. Both seek to make angles 1 and 2 of the NATO triangle compute. Both perceive themselves to be ultimately dependent upon a larger, more powerful, sure, and credible tactical and theater nuclear threat, should they fail.

Because both options are self-confessed servants, it is therefore logical that we should turn our attention from the monkey to the organ grinder to judge the larger issue. Can the entire system, suspended from its nuclear apex, deliver the four types of stability, political reassurance, and simultaneously focused deterrence, in an environment of reductions in number and in readiness of forces?

Let it be clear that this part of the chapter is not about the likely real impact of nuclear weapons upon international relations. That is, in MacGeorge Bundy's precise description, "existential." It cannot be demonstrated on direct evidence. It is wholly passive politically. It sustains no "linkage" to specific tripwires or scenarios. It derives from no formal nuclear strategy, weapon, or deployment. It resides in the ineradicable knowledge that a nuclear device can exist. Its effect is to increase caution in the dealings between already cautious states. The concern here is with "nuclear deterrence"—a time and culture-specific set of ideas, which therefore deserve their inverted commas—and its claims to compute the NATO triangle and to fulfill the three military requirements of the new Europe by deploying an unanswerable capability to prevail.

To reassure as you deter means that the deterrence has to be supple: *strategically specific* and *tactically vague.* Your understanding of and ability to frustrate any form of attack must be transparent. How you would do it operationally must be opaque. NATO's present posture is the reverse of this, principally because of its high nuclear content. All formal nuclear threats, being uniquely and inescapably retaliatory, are inflexible. The special problem with formal European nuclear threats has always been that in trying and failing to buck this aspect of their nature, there is created an inverse ratio between putative credibility in the eyes of the opponent and the increase in specificity intended (paradoxically) to increase that credibility. The NATO predicament is that it is sending the wrong message and that, in order to do so, it compromises its ability to send the right one.

The wrong message is one of retaliatory deterrence: a message of doubt about strategic intentions, always inherently incredible, but the only one that can be made halfway believable by nuclear means. The unwelcome corollary of that doubt is that it has offered the WTO the opportunity to develop some tactical certainty about how to finesse the doctrine by a hard-eyed assessment of NATO's conventional maldeployment, of its nuclear C2, and its likely self-deterrence. In fact (to the astonishment of "common sense"), a nuclear-backed strategy encourages *certainty* that it can be broken at the operational level, but *doubt* about eventual strategic nuclear use. The former point is usually ignored, the latter is usually seen as an advantage that "enhances deterrence" by increasing the opponent's sense of insecurity (a means frequently treated as an end in itself). What we really need to enhance security (which is actually an end) is the *reverse* of these qualities: *strategic certainty* that no form of attack has any reasonable chance of achieving its minimum goals and *tactical and operational level doubt* about exactly how that would be done.

Recognition of the self-deterrence of theater nuclear threats and of the opportunity costs that the present strategy incur led distinguished figures such as the "Gang of Four" to advocate a No First Use policy, the former chief of the British Defense Staff, Field Marshal Lord Carver, to entertain even greater skepticism about any theater nuclear use and the West German government to all intents and purposes to sign off from the battlefield nuclear "rung" of the "ladder" of Flexible Response.[36] But the ties of the heart are strong, and such argument, as during the SNF affair, has continued to be ignored or rejected as tantamount to treason.

The opportunity costs of protecting Thatcher's sense of intellectual security are increasing sharply. It already costs too much. The inability of the present posture and strategy to meet the stability criteria is patent, and thoughtful proponents of the nuclear-laced options would not contend that they do. Rather they argue that the gain of a decisive deterrent outweighs such regrettable loss. What is this gain? What is this power? So erroneous is this deeply rooted belief within the Official Mind that for a moment more we must press on in detail.

The case of the Gang of Four and Lord Carver has been recently buttressed by the first fruits of academic research into theater nuclear C3I, a recondite but undeservedly unfashionable subject.[37] The findings are devastating to any residual case for the present, generally held interpretation of Flexible Response, but are no more than what the paradoxical logic of strategy would predict.

Both Ball and Kelleher are skeptical about the technical capacity of the NATO nuclear complex to offer performance even remotely close to what the strategy prescribes. Ball's technical analysis identifies a tangle of intractable problems. They include the inescapable physical vulnerability of

60 nodal sites (all well known and tabulated in the Appendix of Ball's paper), the destruction of 20 of which could paralyze the C3I system, the communications and command system incompatibilities within the Alliance nations and the depth to which, COMSEC reports have shown, NATO secure communications are penetrated by Soviet offensive SIGINT. This leads to the conclusion that, at "best" (and Kelleher worries about what that might mean), "top-down" release authorization could occur, but that

> . . . the decision-making process is so complex that the most likely outcomes are chaos and paralysis interspersed with precipitous activity . . . Whatever the mechanisms and time periods involved in taking and communicating the dispersal and release decisions, the consequences are likely to be two-fold: first, the distribution of unlocked nuclear weapons to numerous and widely dispersed field units . . . Further, the transition from negative to so-called positive control is unlikely to be reversible in any well-managed fashion. Second, it is likely that, through SIGINT activities and other means, the Soviet authorities would be able to keep abreast of NATO alert, dispersal and release developments.[38]

Kelleher summarizes the politico-military implication. "So long as the role of these weapons is to demonstrate limited intent, the questions, within which limits? defined by who? credible to whom? remain central."[39] Lord Carver's point precisely. In Ball's view, the problems he has identified are insoluble within their own terms. The strategy remains bankrupt, no matter how much money is thrown at it: in fact, the paradoxical logic suggests that to do so would probably only make matters worse.

It is now bankrupt in another sense. The 1989 WINTEX/CIMEX Command Post exercise was the first in several years to exercise the strict implementation of the NATO political guidelines up to and beyond initial nuclear release. What was observable was that the exercise was curtailed and that the West Germans walked out. What is reported is as follows. In the first phase of the exercise, a WTO *blitzkrieg* swept westward. NATO failed to halt it by conventional means and a request for nuclear release was initiated. It called for between 20 and 30 nuclear weapons to be delivered on second echelon and other follow-on forces, deep inside Eastern Europe (exactly the sort of missions that Advanced FOFA is intended to execute). Unusually for such an exercise, the German representative telephoned the real chancellor to seek his view upon the requested theater nuclear package. Köhl suggested, and the other allies agreed, that the package be rather smaller. Being a command procedures exercise, the players now moved on to the next phase. In accordance with the guidelines, a further request for nuclear release was initiated, this time for the employment of a small number of tactical systems (currently the Lance missile, at that very moment in the center of the political storm in NATO) to deliver strikes upon WTO combat forces on West German soil. Again the representative telephoned the chancellor, exercised by the implications of what was now proposed. Chancellor and the German public may not have appreciated the subtle difference between command procedural exercises and an apparently casual game to obliterate Europe. Köhl reportedly replied, "Stop this madness!"

The Germans withdrew. If any politically sentient German had had any doubt of the truth of the slogan "The shorter the range, the deader the German!" or needed evidence of the need to find a different way to strengthen and defend security, or had harbored suspicions about the commitment of the other NATO allies to protect rather than to incinerate them, or had wondered whether there was any sympathy for their wider aspirations of unity, WINTEX/CIMEX 89 appeared to provide the evidence.

"As the paradoxical decline in military utility caused by their own excessive destructive power shows so clearly, nuclear weapons are fully subject to the same logic of strategy that applies to all other weapons," Edward Luttwak soberly observes.[40]

In short, this review of NATO nuclear doctrine, potential employment, and practice is a graphic illustration of how the overwhelming destructiveness of nuclear weapons can only yield, paradoxically, a weak form of deterrence. It does so at the cost of substantial sacrifice of intellectual, diplomatic, and military flexibility. In consequence, by binding the whole system to the nuclear pivot, it is made by far to exceed the culminating point of efficiency. It is rendered incapable of delivering the requisite strategic qualities, neither reassurance, nor stability, nor discriminate deterrence. Such aspects of its conventional dimensions as can contribute to those objectives are hobbled by their nuclear tether. Such support for military restructuring as might exist in the crucial political community, the FRG, may be dissipated and the opportunities to hand to effect a general settlement may be dashed. A more robust security system is required.

The Worst and Best of NOD

This chapter describes a precarious situation. Strong emotions and strongly held beliefs hem in the Official Mind. Much energy is being devoted to a futile attempt to compute angles 1 and 2 of the NATO triangle of problems, and in the process, the triple objectives that should guide the search for security are often obscured. The shielding of current doctrine is an objective that is pursued with relentless and silent passion, of which one consequence is an easy misrepresentation of alternatives, both in substance and in philosophy.

By its structure, this chapter argues implicitly that, whereas it may be temporarily more comfortable to live like this, the time is now past when it is at all wise. Therefore it is important to dissect out the emotional and formal strands of each option. With both of the main options so far considered, it has been suggested that repair and protection of the emotional underpinning has consumed effort that might have been more profitably directed into renovation of the formal structure.

The situation with regard to ideas of "nonoffensive defense" is somewhat different. Here we see views that were conceived in fierce and often partisan emotion, which were born into communities that excoriated the official Western posture as being flagrantly offensive and that, too easily, proffered NOD as a panacea for those ills. Then, in its adolescence, NOD attempted to acquire superficial respectability in the eyes of that official community by dressing in the clothes of mathematical and technical precision, but without benefit of rigorous analysis. The deceit was unmasked. However, in the last few years, little noticed by the Official Mind, which had already been long made up about NOD, alternative defense structures have begun to mature. They have acquired both the reality of mathematical and technical precision and the benefit of a more rigorous formal structure. In short, the progress has been from an emotional toward formal basis and has been in conformity with the paradoxical logic of strategy. The prospect of getting the third angle of the NATO triangle to compute the first is better than the official community might think. But for this to happen, there are two prerequisites. Not only must the inability of current doctrine ever to meet the higher order requirements of the new European security system be faced bravely. The compacted layers of the good and the bad arguments for NOD must be separated and the dross discarded. This is most efficiently done by approaching the best of NOD through the worst.

The following military-technical criticisms are most commonly made by the official community as reasons for abjuring NOD proposals.

1. The proponents of NOD hopelessly misunderstand the significance of new military technologies. Their proposals are in any case narrow in spectrum and open to counter optimization.

In essence, the criticism is that would-be reformers have been far too enthusiastic about the capabilities of precision-guided, rocket-propelled, hollow-charge antitank guided missiles and too swift to announce the demise of the tank. In consequence, their proposals are extremely dangerous. Their technical error has been (just like proponents of FOFA) to embrace the Fallacy of the Last Move. They failed to see that the undoubted efficiency and favorable exchange-ratio of ATGM and tank in ideal one-on-one combat would be swiftly overtaken, for the ATGM could exploit only a narrow technical spectrum, whereas the tank has a broad range of possible countermeasures. The Yom Kippur War did not confirm the ascendancy of PGMs.

This contrast of narrow and broad becomes even more striking when, at the tactical level, are reintroduced the factors of terrain, weather, morale, and luck. Added to the technical response in the action/reaction race between weapon and shield (e.g., reactive armor) and the operational response (e.g., providing advancing armor with mounted infantry screens, machine grenade launchers, more smoke, heavy assault helicopter support, etc.), the presumed advantage of ATGM armed light infantry rapidly begins to evaporate. The exposure of these troops to all the ghastliness of the open battlefield is to be contrasted with the circumstance of tank crews, physically enclosed and, as importantly, psychologically protected by narrow viewing slits and roaring engines.

It is only at the operational level, when mass and momentum enter the calculation, that the full deficiency of the antitank missile "techno-commando" light infantry proposal revealed. For there, all the preceding disadvantages are compounded by the inherent disadvantage of the defense: the need to provide geographical cover. This defense will always fall victim to local inferiority of numbers and victim to counteroptimization of its narrow band of capability by the broad spectrum of the attacker's resources. So the error of the enthusiasts for "defensive defense" was to allow themselves to be seduced by an overpreoccupation with the technical level into ignoring the tactical, operational, and theater levels.

Of course, this criticism is quite correct. The studies of the early 1970s, and those that continue to perpetuate these errors, are justly cut down. But from the perspective of today, this is an attack upon straw men.

2. "Defensive defense" means immobility. Immobility means passivity. Passivity means loss of initiative. Loss of initiative leads to defeat.

General Pollard's belief that the Noddists want to lock him up in a bunker on the Maginot Line is widely shared, and not based upon misunderstanding. Early expressions of NOD (notably the first versions of Horst Afheldt's proposals) did, expressly, stress the desirability of immobility. A countercatechism, "mobility means offensiveness," was a distinctive feature of many NOD proposals. The shorter the range, the slower, the better. Today, the most conspicuous division within the community of advocates is on this question.

Pollard points out that interpretation of a total military posture cannot be made from looking at its component parts. It is perverse to single out a weapon system or a tactic as the general benchmark of intention. A tank in isolation is no more or less offensive or defensive than an ATGM. Furthermore, it is stupid to forego logical innovations because they may appear to be "offensive," if that is what being mobile, fast, or with range is taken to mean. (Skingsley makes the same point with regard to FOFA.)[41] An efficient defense must have offensive capability. As an example of praise-worthy initiative, Pollard points to the recent innovation of operational maneuver groups introduced by Field Marshall Sir Nigel Bagnall while COMNORTHAG. "Bagnall's OMGs" are composed of the reserve divisions of the British and German corps, in conjunction with the Army Group reserve, operating under AG command well back from the FEBA to counter breakthroughs or to exploit opportunities for localized offensive action, which would regain initiative for the defender.

Pollard is quite right. The "mobility is offensive" Noddists commit a blundering category mistake: they introduce strategic judgments, such as what is to be deemed "offensive" or "provocative," and address them at the *technical* level. A putative "nonoffensive" quality is unobtainable when prescribed in technical terms. Again, from the perspective of the late 1980s, this is a misdirected attack.

Refinement of thinking about mobility is the best indicator of the move from emotional to formal drive, from the inferior to the better work, in the alternative defense community. The crude equation of mobility with offensiveness has been discarded by the more sophisticated proposals. It has been replaced with the more realistic and useful distinction between *tactical mobility*, essential for stalwart defense as General Pollard explains, and *strategic combined arms mobility*, which it is a central arms-control objective in CFE to reduce and eventually to remove. Thus we see the functions of Bagnall's OMGs portrayed in the alternative literature first as those of the "Sword" forces in General Jochen Löser's 1981 proposals and more recently, as Albrecht von Müller's "conditional counter-offensive capability" and "close interdiction capability."[42] The greatest detail on this aspect has so far come from the SAS (*Studiengruppe Alternative Sicherheitspolitik*) Group, which prescribes 30 battalions of heavy and 40 of light mechanized armored forces deployed in two "net" zones, each 36 kilometers deep behind a 14-kilometer "fire zone." They behave like spiders in the web.[43]

In the most recent German alternative proposals, there is much closer convergence upon questions of tactics and equipments with the more imaginative aspects of official thinking than the official community apparently yet knows. At the operational level, the necessary synergism between ATGM infantry and other reforms, notably the preparation of terrain, which might remedy the failings of a narrow spectrum defense are well understood. Current reform proposals are built upon a recognition of the dangers of counteroptimization (and indeed, although it is not often stated, in the minds of some advocates, upon the conclusion, noted above, that nuclear-backed forward defense is conspicuously open to it). Therefore at their heart lies the integration and redistribution of armored and light mobile forces, of force multipliers derived from the terrain, of SP tube artillery, of MLRS, and of new forms of cargo rocket, especially in conjunction with a much extended inventory of passive ammunition.

The recent development of new mine warfare potentials (stochastic mines, jumping mines, etc.), notably by the MBB Company of Munich, does not imply a "Maginot Line" because it is not a wholly passive and immobile defense.[44] The NOD reform is *not* a narrow spectrum proposal, based on a few military technologies; and unlike FOFA, it does not call for exotic, new technologies, with multiple, fragile, and therefore vulnerable data and command links. It is not a "technological fix." In fact, it is precisely the creative synergism and the fact that none of the components are pushed beyond the culminating point of their military efficiency that makes this type of reform prospectively far more robust than what we have at present.

The goal of improving the four types of stability prescribed earlier increasingly replaces the search for a platonic and chimeric quality of "non-offensiveness." Seminal to this is the best of the conceptual and modeling work in the Federal Republic, which has been done by the Operations Research group at Universität der Bundeswehr, Neubiberg, near Munich. An early and constant feature of their work has been the comparative analysis of the different proposals, including current doctrine, according to Professor Hüber's "defense efficiency hypothesis." This was done in combat modeling of a first echelon engagement first at battalion scale, now at larger units of engagement. With due caution, this work was among the first outside the community of advocates to find military virtue in aspects of their proposals. Potentially, they offered ways to improve both "confrontation stability" and "defense dominance."[45] At the same time, the group recognized that "political crisis stability" and "arms race stability" were essential, and so paid considerable attention to technological provocation and to robustness against pre-emption.[46]

Addressing directly the uncomfortable relationship between force planning and arms control was early seen to be vital. UBw(M) papers were again among the first to alert opinion to the considerable potential of restructuring proposals to maintain stability under a regime of cuts.[47] Hüber has also not avoided that thorniest of issues in dispute, the degree to which strategic stability in Europe depends upon deep strike and nuclear systems. Some of the most interesting of the UBw(M) work has been that which relates to this question. Of course, it does not offer strategic assessment. Understandably, Hüber will not extrapolate beyond his combat modeling. But from this, he will sustain the minimalist position (now widely held in the OR community) that mutual restructuring of both alliances toward defensivity in Europe by the removal of the assets required for strategic mobility could be managed without reference to the nuclear dimension for compensation.[48] Although only mentioned in passing in his work, like that of the other researches mentioned (and it will not have been lost upon the reader that they have all been West German), it is predicated upon a given context of the erosion of belief in nuclear deterrence and support for the emplacement of nuclear weapons in the FRG. Remember WINTEX/CIMEX 89.

Increasingly, as they are refined, and as they converge with the most creative thinking within the official community, the detailed German NOD proposals are being offered and accepted á la carte. But should they be? Ideally, they should not. Failure to follow through the logic of the argument has two consequences, both grievous. The first it entrapment. Proponents find themselves trapped in the lower military-technical levels of strategic analysis from which, in their desire to engage the higher strategic levels, they make the category mistakes for which Professor Tromp and the present author have criticized them. Secondly, and in consequence, the likely consequences of adoption of the best of NOD are not plainly stated, as they should be.

The first consequence of achieving greater crisis stability through restructuring at present force levels is that it will indeed harmonize with the objective of lower numbers of conventional forces in forthcoming conventional arms control. This is sharply different from the present posture, where reductions will shorten the time to second breakthrough, thus greatly increase the bonus for pre-emption and therefore reduce military efficiency. Furthermore, the lust to modernize below whatever ceilings are agreed, can be channeled productively. This is a major gain for the first-order criteria of collective security.

The second consequence is paradoxical, but consistent with the logic of strategy. It is that a self-contained defense by denial can, unlike the nuclear-backed hybrid, plan realistically how to thwart the full spectrum of Soviet *blitzkrieg*. This is for two reasons.

The first reason is a further insight of the paradoxical logic of strategy: *Less than absolute destruction may be operationally more desirable than absolute destruction.* The minimum Soviet requirement under plans for *blitzkrieg*, abundantly attested in literature and in exercises, is for the maintenance of momentum. This is to achieve swiftly the objective of the first phase of world war, to reach and secure the Atlantic perimeter. It must be done before NATO can recover from paralysis and so as to release forces for the northern TVD. Therefore imposition of delay pays NATO *disproportionate* dividends. A mobility kill in WTO tracked armored forces is as strategically efficient as total destruction, cheaper and easier to inflict, and less easy to resist. Just such a capability comes from the new barrier and mine components of the reform. The Hungarian guerrillas in 1956 found soap and oil slicks and Molotov cocktails to be efficient also. It also provides a perspective within which the breathless debate about the putative relative advantage of ATGM and reactive armor may be seen to be less central than many in the armed forces currently think.[49]

Such ideas take some time to catch on, even with the degree of support that some parts of them now have, especially in West Germany. Why? General Sir Hugh Beach suggests that for professional military, such "small is beautiful" ideas ". . . have a profound suspicion of crankiness. There is a comfortable and reassuring solidity about a 50-ton tank, a self-propelled gun, an armored personnel carrier."[50] He is surely right. For such minds, contemplation of the paradoxical logic of strategy is a vital corrective:

> . . . once the dynamic consequences of the logic are understood, the exercise of restraint in pursuing success in war and peace, in the conduct of an offensive or in the building of weapons, will no longer depend on vague instincts of moderation but can instead be sustained by a compelling rationale as culminating points are approached.[51]

The second reason why a self-contained defense by denial can plan realistically how to thwart the Soviet *blitzkrieg* is that by *nonnuclear reform*, a credible solution to the threat of Soviet *nuclear* use in theater can, at last, be offered. This is no paradox either. As Luttwak remarks, ". . . the effectiveness of suasion would be circumscribed by motive . . ."[52] He postulates that especially a Soviet attack driven by despair might be not amenable to nuclear dissuasion. Much evidence suggests that Soviet operational planning for *blitzkrieg* was predicated upon a presumption of NATO's theater nuclear self-deterrence.

However, the sting of Soviet nuclear theater threats, which cannot be resisted by counterthreat, may, however, be drawn by removal of their rationale. If by dispersed formation there are no decisive, lucrative targets; if the military gain of any nuclear use—for example, to blast through defensive belts—could be more speedily nullified than sealed-up forces could exploit it, then in the hierarchy of risks, a better bet has been found. The Soviets can be left to calculate the higher order political and strategic costs of nuclear use, and their self-interest, not their fear, relied upon. In the next section we note that there is evidence of a debate about "nuclear sufficiency" in theater that has been running in the Soviet General Staff for some time.

Therefore the third consequence is that the reform strategy does not call for any articulated nuclear "linkage," any promises to do incredible things in highly unlikely circumstances. The "Third Zero" is to be welcomed, not dreaded. This is fortunate because, as commentators as varied as the authors of the *Discriminate Deterrence* report, Paul Kennedy, John Palmer, and David Calleo have recently suggested, the likelihood of the United States being prepared to make such statements (or indeed to provide the rest of the present level of military commitment to Europe indefinitely) is diminishing.[55]

Therefore, it appears that just in the nick of time, we can indeed legitimately strike the shackles from the interloper NOD on the grounds that it better conforms with the real logic of strategy than any other option to hand. We can allow it to help angle 3 compute angle 1 of NATO's triangular problem; we can even call it a prudent evolution of Flexible Response, and thus preserve appearances in front of the neighbors. On the bankruptcy of

formal nuclear strategy, we shall maintain a discreet silence, out of kind-
ness, just as a considerate woman would if she found out that Casanova was
impotent.

3. Why none of these options would work, or are likely to be adopted in
   full, but why the best of NOD is nonetheless worth pursuing.

We could do this, and perhaps should, but as we do we should be entirely
clearsighted. So there are two other little secrets that we must know and
keep. None of these options would be likely to work in practice for a quite
different practical reason, so far not mentioned; nor, knowing it, will the
effort be made to rectify the situation. But, as the next section suggests,
fortunately that probably doesn't matter for NATO/WTO relations. In fact,
touching an underlying thesis of this chapter, it may be really quite a relief
if it turns out that there are no conclusive deterrents, neither of retaliation
nor of denial, stalking the new Europe; only winded, superannuated boxers.
It may take people some little time to grow comfortable with that realization;
and so it is may be good that the process of winding down will take some
time.

The cleverest practicing historian of war writing in English, Martin van
Creveld, has returned to the field of his first triumph. *Supplying War* was
a virtuoso demonstration, sweeping across modern history, of the centrality
of logistics to victory. From it we have learned that Hitler's defeat in Operation Barbarossa owed more to frozen locomotive steam pipes, problems
of rail gauge change, lack of truck antifreeze, and incompatibility of Russian
fuel and German engines than to the tactics of the Panzers. We learned to
admire Napoleon for revolutionizing the supply of armies.[54] That book ended
with Patton's famous advance across France, and the less celebrated supply
trucks of the "Red Ball Express," which enabled him to do it. What has
happened since?[55]

Since the end of World War II, of the four principal technologies of warfare, combat mobility and armor have improved, but the abilities to allocate
and deliver fire and to process information have been revolutionized by complex new machines. As part of and on top of this, units have hugely more
destructive potentiality as a consequence of being far more extensively and
intensively equipped than in 1945. There are more guns, vehicles, radios,
and so on per man, and new systems like missiles and helicopters, which
were not there a generation ago. The consequences of this for the ability to
sustain combat have not been fully appreciated. In one field of applicable
military technology, viz. mass transport, there has been very little improvement. The principal means of transport in the theater of operation has not
changed. It remains the 4X4 or All-Wheel drive truck. Neither tracked vehicles nor aircraft nor helicopters are remotely able to substitute for it.

Modern warfare consists essentially of machines moving around firing metal into the air. The requirements for ammunition and for Petroleum Oil Lubricant (POL) compose, by volume, about 70 percent of combat supplies. The needs of men (food, medical supplies, etc.) compose less than 10 percent. The growth in the appetites of these machines for consumables, under the influence of technology, which has given more powerful engines, heavier, faster firing artillery, automatic weapons of all descriptions, has been gigantic. Even in a miniature modern war, they starved the men. Describing his experience in command of the British landing at San Carlos in the Falkland Islands, Major General (then Brigadier) Julian Thompson records how the demands of the artillery upon the logistics meant that one battalion of 3 Commando Brigade was without rations for 72 hours. "By the end of the first day [of the landing]," he writes, "it became clear that logistics would dictate the pace and texture of the battle; and they did so to the end."[56] In 1941–42 a well-equipped German Panzer division required about 350 tons of supplies per day. The more lavishly equipped Allies in France after D-Day required 650–700 tons/day per division. By the Korean War, the requirement was over 1,000 tons/day. No one has calculated the figure since. There are, van Creveld reports, guesses of about 1,500 tons/day for a modern NATO division on the central front.[57]

It has been argued against this that technology has actually reduced the requirement for ammunition tonnages. You need fewer precision guidance munitions (PGMs), because they hit more reliably. But this is, van Creveld observes, simple-minded. The evidence from the Middle East wars has been that the presence, or threat of presence, of PGMs greatly *increases* ammunition expenditures: the machine gunner atop an APV responds to the threat by saturating with suppressing fire every possible bush or building that might conceal a PGM-armed soldier.

It is not just volume that is the problem but the complexity of modern combat supplies. In Vietnam, the semiautomated inventory control broke down. Units would send soldiers in jeeps to scour at random the supply dumps at Da Nang to find what they needed. Computers can deal with the mind-boggling numbers of parts, but they can break down, or be sabotaged, and they are in any event, quite unable to improvise. Van Creveld suspects that put to the test, catastrophic failure of inventory control would be more rather than less likely in an European war. Even at San Carlos Water, Brigadier Thompson's people had problems keeping track of where things were when they had to off-load under pressure.

Van Creveld suggests that all the indicators are that no one, probably, could supply the modern European battlefield successfully for very long. The Soviet tank armies might roll, but before they reached they Rhine, they would stop, and they would never make it to the Channel, for reasons of logistic failure. And all this presumes that there is any modern economy that could sustain the burden of generating in wartime such gigantic volumes of goods for war purposes. As every terrorist threat demonstrates, modern industrial societies are extremely fragile. Europe, after all, will have no lights, no water, no sewage treatment, no trains, no food distribution, breached chemical works along the Rhine and the Danube belching toxic clouds, pretty soon after the start of any form of hostilities, and no one seriously believes that there is sufficient transatlantic shipping to supply the battlefield reliably from the United States. The truth, van Creveld concludes, is that modern, strategically mobile warfare as a whole has, in volume and complexity, exceeded the culminating point of efficiency. It probably just can't be done.[58]

Is the same true for a defensively restructured posture? Probably so. In place of the retaliatory threat of nuclear oblivion, NOD substitutes the older, also unpalatable but more credible threat of a long war. Yet a cardinal technical failing of all the NOD proposals is that they have so far paid no attention to logistics. The SAS model, the most detailed to date, is a "snapshot," not a movie. Unterseher freely admits that his group hasn't considered logistics.[59] Equally, with the intensely Eurocentric—in fact Germanocentric—focus of NOD work, minimal attention has been paid to the maritime implications of the need for huge volumes of ammunition, POL, and foodstuffs to be shipped to Europe.[60] Again, the suspicion must be that it can't be done. More certainly one may predict that the huge new expenditures that would be required to procure, maintain, store, and ship modern warstocks in peacetime on a sufficient scale will not be forthcoming. Nor should they be, as Europe and the superpowers attempt to move away from the abnormalities of high readiness and high alert armed confrontation of the last 40 years. It would be self-defeating.

Does it matter? Again, probably not. Unterseher's answer is politically acute. We don't need the movie at this time; we do need the snapshot. We must have ways of traveling hopefully, not of guaranteeing arrival at a destination that we hope we shall never, ever have to reach. That means adopting military postures that conform with the three objectives set out in the previous section and especially that maintain and increase stability and confidence as force levels and readiness begin to fall. The fact that NOD is probably as little able as any other military strategy to achieve its mission in battle in Europe in the 1990s is neither here nor there. One further attraction of NOD is that unlike options 1 and 2, the troops would not become downhearted if told so, because a coherent alternative mission in support of disarmament can be described in detail.

However, genuine disarmament is a two-handed game. Since the Budapest Appeal, the WTO has said with strengthening voice that it agrees. Astounding things have been said. The WTO, and especially Soviet spokesmen, say that they see defensive restructuring as an integral part of the game. What does this mean? In the NATO Official Mind, as well as in any reasonable West European mind, fueled by years of suspicion and, even if sympathetic to the Second Russian Revolution, fearful of what might follow its failure, there is reluctance to take such sweeping conversion at face value. Diplomats complain that "automatic Da!" is as difficult to deal with as was "automatic Nyet!" If everything is said to be open for discussion, frequently the practical result is that nothing can be. What does the WTO mean when it says that it espouses ideas of defensive restructuring? Do actions confirm the words? So long as answers to these questions remain opaque, so long will hesitation continue in the West about facing the situation described so far in this chapter and taking the actions that are indicated. Therefore in the final section, and deliberately left until last, we interrogate Banquo's Ghost seated at the Feast of NOD.

"NONOFFENSIVE DEFENSE" IN THE EAST: FROM ICE TO WATER

In 1922 the USSR (as today) was in desperate need of tranquility in its external relations and of an increase in foreign trade in order to rebuild its shattered economy.[61] Accordingly, the Lenin leadership decided secretly to adopt a deceitful posture of pretending to have moderated its international revolutionary doctrine in order to buy a *peredyshka* (breathing space) in which to recruit its strength. Delegates to the 1922 International Economic Conference at Genoa were required to have an excellent knowledge of the arguments of "bourgeois-pacifist" books, like Keynes' *Economic Consequences of the Peace*, to play back to the Western liberals who, thus deceived, would put pressure on Western conservatives to relax pressure on the new Soviet state, which they had just recently failed to destroy by intervention and by support of the Whites in the civil war. This objective could be achieved only by the "indirect approach."

There are some striking, superficial similarities with the career of "new thinking" during the last few years. Writing in *The Times* of London at a moment when the Western world was gazing star-struck and delighted at the spectacle of "Moscow Nights" being played for the Gorbachevs in the White House during the Washington Summit, Lord Chalfont warned the giddy masses.

> In the deceptive glow from the artificial lights of summitry, the West ought to contemplate a sombre proposition . . . If [Gorbachev] succeeds, we shall be faced with an adversary whose basic foreign policy remains substantially unchanged, who will be better placed to pursue that foreign policy by the threat of armed force and who will be immeasurably more effective and self-confident.[62]

Many of the central tenets of "new thinking" are, indeed, recycled Western liberal ideas as expressed, for example, by the Palme Commission in 1982. The "Common European House" of Gorbachev's vision does not appear to boast any revolutionary new architecture. He would seem to be content to live in an apartment of the social-democratic version of contemporary Western Europe. He says simply that it combines ". . . necessity with opportunity."[63] Are these views to be interpreted as Chalfont thinks?

In the case of "nonoffensive defense," there has been a veritable deluge of accessible publication by the Academy of Sciences Institute researchers most involved with this issue—Arbatov Jnr, Kokoshin, and Co.—which, on close reading tells rather little about new Soviet ideas but appears to recycle the Western liberal ideas about NOD.[64] Before 1987, it was believed that these researchers had little influence with the General Staff, which is just what would expected if their writings were only for external consumption. Indeed, at one point in that year, Gorbachev's close advisers approached directly some prominent Western proponents of restructuring and asked them to prepare a memorandum for Gorbachev, elements of which were later to be seen, coincidentally or not, in subsequent Soviet pronouncements.[65] In the most concentrated statement on threat perceptions and military restructuring by the institute researchers[66] as well as in the May 1987 WTO call for a general reduction of weapons to the level where no one would be able to conduct offensive operations and implicitly by the WTO's acceptance of the essentials of the Western draft for the CFE Mandate, the Western perception of the aggressive posture of WTO forces was also recycled. Should this cause us to relax or to tense our trigger fingers?

Western analysts believe that growing importance should be attributed to the newly established Scientific Coordination Centre in the Soviet Foreign Ministry. It seems to be what its name says. It seems to be a clearinghouse, coordinating policy. It seems to fit the picture. The thought that this may all be an elaborate and very successful repetition of 1922 can be further strengthened when the mask appears to slip. This author recalls well when in 1984 proposals for unilateral Western adoption of "nonoffensive defense" were first ventilated before some relatively low ranking and silent Soviets. When, eventually, they spoke, the reaction was extremely hostile. The prepared response was that these proposals appeared likely to strengthen NATO's ability to resist the onslaught of the WTO "offensive defense" and therefore were certainly not "reassuring" as the proponents suggested. On the contrary, they were perceived as extremely aggressive!

The differences between 1922 and now are far greater than any similarities. Then, revolutionary zeal was high, immediately after a remarkable victory in war. Stalin had not yet murdered the Old Bolsheviks. Today Marxist-Leninist ideology is under widespread internal attack after 40-odd years of peace. This internal debate is occurring in a world where the new information technology revolution has denied any closed leadership the assurance of being able to lock up their peoples' awareness for long. In fact, in the Soviet case, the relationship to the media is central to the strategy of the *perestroikists*. They know that their enemies with wolves' eyes are in the shadows and they seek security in the light of the flame of *glasnost*.

Gorbachev and his supporters are attempting the most difficult sort of social revolution. Having initiated a revolution from above in a country where the reward for initiative has usually been painful, or terminal, they are trying to turn it into a revolution from below. They know that such a maneuver tends to disintegrate at the moment when the top has to relax its grip to give power to the bottom. Knowing it, they are trying to use the new opportunities of the present to avert it. Gorbachev stated in an unscheduled speech at Krasnodar in September 1986 (which, MccGwire argues, was in many ways the text for the critical period when the present course of *perestroika* was hammered out and agreed, between the plenum of January and that of June 1987) that democratization was an inescapable prerequisite; and therefore *glasnost* had to precede the new revolution to give it a fighting chance. The most irreversible thing that the *perestroikists* can do is to flood the country with photocopiers and micro computers.

Furthermore, most of the basic ideas of "new thinking" were not adventitiously picked up in the last two years. They have been in discussion since at least the mid-1970s, had Westerners both known to and cared to look.[67] MccGwire believes that the debate about military objectives within the General Staff also precedes the arrival of Gorbachev, starting with Andropov's September 1983 declaration.[68] Although we must walk carefully, we must also learn to distinguish shadows and substance. With such a legacy of mistrust, it is not easy to do so. However, as the Pentagon Sovietologist Phil Petersen concludes (deliberately paraphrasing Kokoshin), we have to see everything as it really is, without making illusory assumptions.[69]

MccGwire isolates four such assumptions commonly made in the West as we look at and try to understand Soviet military power.

1. The assumption that the aggressive posture of the offensive defense prescribed by the doctrine, which held from 1966 to May 1987, was more than a capabilities threat, reflecting a genuine urge to invade.
2. The assumption that the Soviets confuse the military-technical and military-political levels as comprehensively as we do.
3. The assumption, following from no. 2, that the rather systematic process of decision making, whereby the Soviet authorities proceed from analysis to a "decision in principle" to enunciation and implementation, is too neat to be plausible.

4. The assumption that what appears to be revolutionary military change is revolutionary military change.

If assumption 1 is held, it is understandable that Western observers find it hard to believe the new statements on doctrine at face value and easy to remember the Genoa Conference of 1922, and Lord Chalfont. But if, as MccGwire demonstrated to this author's satisfaction in his earlier book, the objective has been one of securing national defense throughout, the present changes can be seen to be more evolutionary than revolutionary: a debate about changing means, not about ends. If assumptions 2 and 3 are held, we are fated never to understand how the Soviets go about it.

Assumption 4 fails to understand the process of change. MccGwire offers a helpful analogy. As ice is gradually warmed, the temperature rises steadily, although the ice seems unchanged: an evolutionary process. But above the freezing point, quickly the ice melts and becomes water, something visibly quite different: an apparently revolutionary consequence of an evolutionary change. By dropping the four common assumptions, we may be able to see a defensibly credible outline of where and why ideas about "defensivity" fit.

The first response to Western ideas about improving defensive capability was at the military-technical level.[70] For a long time, Soviet military planners had been deeply concerned about an aspect of NATO's capabilities that NATO itself constantly belittled—its firepower (whereas a prime asset of the spearhead of the attack was NATO's real weakness—its maldeployment, imposed for political reasons). The planners saw that the ability of WTO forces to absorb the damage that NATO could inflict upon the flanks of the advancing columns of the *blitzkrieg* was poor. They also greatly feared the ability of NATO air power to conduct offensive counter air and other deep interdiction missions. In the new defense-oriented munitions and tactics, they saw a way of strengthening the flanks. It is perfectly legitimate to invert General Pollard. A stout defensive capability is essential for a robust offensive. By encouraging the rhetoric about "mobility is offensive," they saw a way to begin to get at NATO air power by diplomatic *maskirovka*.

At the same time, in the mid-1980s, it was plain that change was in the air. The General Staff responded as Russian generals have done many times since Napoleon, by reshaping formation from army to corps, by restructuring divisions, by tightening command and control within formations, and by creating extra, autonomous theater C3I capability (for example, south of Lake Baikal for the southern TVD). It is a process similar to battening down a ship whose captain knows that dirty weather and rough seas are ahead but who cannot predict the precise direction or nature of the storm.

Then in 1983–84 came the first of the major reassessments at the military-political level. It conforms with the correct application of the paradoxical logic of strategy. Quite apart from the influence of memories from the Great Patriotic War, an offensive defense is *inevitably* prescribed by the geopolitical fact of the shape of Europe and by a requirement to prepare not to lose a world war, should it occur. Yet it was accepted that this military-technical prescription of the previous 15 years was jeopardizing the higher order objectives of national security, overreaching the culminating point of efficiency. Adoption of the posture had scared the pants off the Western Europeans. It legitimized continuing high levels of funding for the Western military-industrial complex. It frustrated the diplomatic objective of increasing conviviality in the "Common European House." It increased mistrust. In fact it made war more likely. Furthermore, it imposed a crippling burden upon the Soviet economy and left the Soviet General Staff with too much influence upon the conduct of the domestic economy. In May 1986 in a Foreign Ministry meeting, Gorbachev is reported to have said that the correlation of forces was untenable.

However, these military-political arguments did not lead to a head-on confrontation between reformers and the military. The General Staff had and has its own reasons for welcoming the prospect of substantial cuts in standing forces. Quite apart from the difficulties to which van Creveld draws attention, of maintaining and supplying the volume of force required for the strategic offensive, it had become seriously worried that the minimum tempo of operation required for *blitzkrieg* could not be sustained with the poor quality of junior officers and apathetic soldiers in an army increasingly riven by interethnic friction and frustrated by poor living standards. The training and exercise problems of introducing more potent versions of current systems to such soldiers only increased the problem. To maintain tempo on a future battlefield, the *lowest* force density consistent with cohesion is required. There were other military reasons also. The USSR is badly deficient in the technical skills and manufacturing resources needed for the next generation of weaponry, with its high premium on C3I. If compelled to maintain and equip current levels of present generation forces, this would be both an opportunity cost choice against freedom to acquire the new generation in the next procurement cycles and a self-defeating move if its development required access to a strengthened new technology base in the economy, the creation of which was predicated upon improved access to external know-how.[71]

This self-interested military-technical argument in favor of deep cuts (the so-called lean and mean argument) has been much rehearsed in the Western military establishment. However (illusory assumption no. 2 above), it has failed to see that, whereas it may be correct in its own terms, it does not follow that Soviet policy is being driven at this level just because an equivalent case might have that political power in the West. Nevertheless, while "lean and mean," with all the resonances of 1922, must be kept in proportion, it does draw attention to the delicacy with which Gorbachev has to handle his military establishment. He would find it extremely difficult to proceed if his reforms and cuts imposed for higher level reasons could not be made to conform to what the General Staff believed to be a competent military plan.

It is beyond the brief of this chapter to do more than mention the very considerable and significant changes in personnel that Gorbachev has brought about at the top of the armed forces, and the fact that a reform of the General Staff itself is now under study. It is not necessary to explain Matthias Rust's flight by conspiracy theory to see that Gorbachev has been extremely adept in this regard![72] The competent Western students of these matters seem to agree that (1) the General Staff will do its best to accommodate Gorbachev's demands, (2) that he is thoroughly unpopular among the middle-ranking and junior officer corps, (3) that on past performance, he prefers to let the military have time to work out their plan, but that (4) if they don't, or take too long, he will override them, as apparently he overrode Marshal Akhromeyev's opposititon to unilateral force reductions in December 1988, leading, it was thought, to a strategic bout of illness, which was given as the marshal's reason for early retirement.

In 1984 the military-technical objective of not losing a world war was replaced with one of averting that remote contingency. The consequence of this doctrinal adjustment became diplomatically active in the June 1986 Budapest Appeal, although already, in his "nuclear disarmament by the year 2000" statement of January 15, 1986, Gorbachev himself was already pegging out a route well beyond that step. Meanwhile, the military-technical consequences of this change in objectives were being digested by the General Staff. The first was to reduce markedly the gross force requirements, the objective set being now that of traveling hopefully, not of guaranteeing arrival. The impact of this was not observed in deployments, exercises, or production levels during 1986–87. That was reserved for 1988. But at this time a second consequence was seen. There was evidence noticed of increased interest in defensive battles, notably the Kursk Strategic Battle of 1943, in Soviet operations analysis.

It now appears that during the summer and autumn of 1986, Gorbachev and his closest advisers concluded that the state of the domestic political economy (including the moral degeneration of peoples' spirit) was so parlous that decisive steps, including confrontation and vanquishing of opponents, could not be avoided. *Perestroika* was going to be more than just fixing the economy. It was going to be revolution.[73] The January 1987 plenum of the Central Committee formally added the policy of *demokratizatsiya* to make the trinity with *glasnost* and *perestroika*. In the ensuing months, before the June plenum ratified the strategy for economic reform, press liberalization moved forward, the slow release of the true history of Stalin began, Sakharov returned to Moscow, American seismologists were allowed to reside near Semipalatinsk to monitor nuclear tests, and the INF issue was detached from the rest of the Reykjavik "package," thus opening the way for the signing of the INF Treaty at the Washington Summit in December 1987.

Sakharov had returned to public life in February 1987 on the occasion of the Moscow Forum. At it, in the keynote speech before the assembled scientists, scholars and *glitterati* of the Western world, Gorbachev made a further powerful speech on the powerlessness of any individual nation-state to secure itself and on the need for supranational cooperation. Its theme was quite consistent with that of the Warsaw Pact Berlin statement in May. This made public the second major reassessment at the military-political level, which was to the effect that world war could be averted only by political means. Military power, expressed in deterrent threats, had no competence any longer and so the military should cease to prepare for the eventuality while maintaining and strengthening defensive preparations against the "harsh reality" of the West's military preparations.[74] From this moment, "reasonable sufficiency" became the watchword, and defining it the name of the game.

During 1987, the speed of preparation within the General Staff seems to have increased. British Sovietologist Roy Allason believes that from early that year, the status and influence of the civilian analysts changed. Their views on the value of NOD proposals, their advocacy of the adoption of criteria of strategic stability to be obtained without nuclear reference, and their warnings on the damaging higher order consequences of persisting with what the West perceived as aggressive postures gained official endorsement, for example, from Major General Larionov. On their own admission, people in the General Staff began to return their telephone calls and to attend their seminars, whereas previously they had not.

In the early 1980s (again, well before Gorbachev's arrival), the idea of "nuclear sufficiency" had emerged from the General Staff. It appears to be fairly close to Western concepts of "existential" deterrence and, in the view of some but by no means all Western Sovietologists, had already weakened the General Staff's attachment to its theater nuclear systems. This may explain why it was prepared to accept so asymmetric a deal under the INF Treaty and to lose not only the SS-20 but also deployed and operationally integrated tactical missile systems. As has been seen, any formal (as distinct from emotional) requirement for theater nuclear potential is removed with NOD. Therefore, since the removal of NATO's tactical and theater nuclear systems was both a mandatory operational requirement of the *blitzkrieg* strategy and a significant political component of normalizing relations within the Common European House under the new objectives, Allason believes that it is possible that NOD ideas found a sympathetic audience in the General

Staff after 1987 for this reason also. It was by then evident to the staff that Gorbachev was going to press for deep cuts in strategic nuclear systems, and although Allason does not believe that the constituency for total strategic denuclearization is large, he thinks that those now willing to entertain bilateral European denuclearization probably consitute a majority.[75]

Allason credits the analysts with considerable influence in changing the traditional threat perceptions of the General Staff and with bringing the analysis of interests as well as capabilities more centrally into its thinking. At the same time, Marshal Akhromeyev's appointment upon retirement as chief of the Soviet General Staff to be Gorbachev's personal military adviser was a brilliant stroke. It came just at a moment when the Western *tricoteuses* were prepared to hear the marshal's retirement as a rumble of the approaching tumbrel bearing Gorbachev to the guillotine of coup d'état. In fact, Akhromeyev's new post makes a gamekeeper out of a poacher in Gorbachev's struggle to curb the power of the General Staff. Akhromeyev has been there: He knows the tricks of the tundra. The General Staff's monopoly on information—its most important monopoly—has been broken.[76]

The interest in defensive operations analysis has further increased. The opening phase of the German invasion in 1941 (when Soviet defenses failed so catastrophically) was the Topic for Study set in the Group of Soviet Forces in Germany in 1988. In addition to Kursk, the repulse of the Japanese incursion from Manchuria into Mongolia in 1939 at Khalkin-Gol is held up as a model of defensive thinking.

Both battles had common features in the use of forward fortified areas to lure and to erode the enemy's attack, followed by major counteroffensives to break the weakened enemy. At Khalkin-Gol, for political reasons, the counterattack stopped at the border. The attraction of such a model of operation to the General Staff could be great. The convergence with some of the Western proposals for NOD is self-evident.

Assessing what this might translate into in Europe, Chris Donnelly of the Soviet Studies Research Centre at the Royal Military Academy, Sandhurst, suggests that we should now expect to see really significant changes in Soviet structure and deployment. In his view (at the time), it might have meant the maintenance of border covering forces by the other members of the WTO and the creation of "fortified regions," terrain fortifications saturated with antiaircraft, antitank, and rapid fire antipersonnel weapons and manned by Soviet and East German troops. Soviet armored divisions might be drawn back from the border and reorganized into counterattack units in the rear (the SAS "Spider" force) and C2 might be reorganized to cascade more initiative down from military Theater to District commanders.

Putting himself into the General Staff's boots, Donnelly perceives advantages for the WTO from such a reorganization. At the military-technical level, they include the ability to reorganize combat forces in ways that are better able to keep pace with new tactics and technology and that permit the General Staff to retain its offensive capability, qualitatively enhanced by the overall reduction in numbers. It also renders WTO forces less vulnerable to new technology Western weapons by virtue of being able to fortify and to deploy in depth. At the military-political level, the advantages are the fact that such deployment would appear much less provocative to the West and that it could probably be achieved with about 50 percent of existing forces levels.

If that is so, then the 10 percent unilateral cuts that amazed, delighted, and confused simultaneously different interested constituencies, announced by Gorbachev in the speech at the UN on December 7, 1988, taken together with Foreign Minister Shevardnadze's proposal at the opening of the CFE talks in Vienna, fall into both their military-technical and military-political contexts. The Shevardnadze proposal is for cuts in six categories of weapon essential for strategic offensive action to levels 10–15 percent below current NATO levels, with 25 percent further cuts and restructuring of remaining units to follow in subsequent phases. Put together, the projected target force level indeed looks like about 50 percent of 1988 deployments. It is frankly inconceivable that such dramatic reductions could be proposed unless the General Staff has approved a plan for reconfiguration. We may expect it to make concrete the doctrinal statements about defensive orientation of recent years.

This is especially so when it is recalled that the WTO is really burning its boats with respect to preserving the capability for offensive defense in Europe. The unilateral cuts were allocated in militarily meaningful ways from that point of view. The removal of assault landing and assault bridging forces from GSFG, the withdrawal of six divisions, and removal of late model tanks from forward assigned units led most NATO staff assessments of the impact of the December 7, 1988 cuts to agree that they removed the threat of the "bolt from the blue" short warning attack (always assuming that you believed it to be a real possibility in the first place).[77] Still unpublished Western civilian analysis of the likely impact of effecting the Shevardnadze reductions concludes that they would remove the possibility of a successful WTO attack after full mobilization.

In plain words, the Soviets are giving up the ability to evict the United States from the European continent in the event of a war, the thing that is the *sine qua non* of not losing a world war. In return, they have received the means, in the form of surplus forces, to be able to make unilateral cuts on a sufficient scale to prime the pump of subsequent multilateral reductions. They have, in Les Aspin's memorable phrase, "left us looking like a bunch of bookkeepers, while they have a vision!"[78]

Where is this all intended to go? Another British Sovietologist, Peter Frank, suggests that the answer is plain enough if we allow ourselves to believe what we hear.[79] Gorbachev's statement of January 15, 1986 in which he called for universal nuclear disarmament by the year 2000 and the attainment of the highest world level of social productivity was greeted with incredulity in the West. Lord Chalfont called it "foolish chatter." Thatcher repeated not for the first time that nuclear deterrence is eternal and the apparent utopianism of it struck the Western peace movements dumb: they hesitated to endorse it since it might make them seem too utopian.

However, Frank argues, Gorbachev has a track record of planning carefully what he says, when and where he says it, and he has not forgotten the January 15, 1986 speech. Speaking to the Soviet people on television to explain the INF Treaty, he said, "The philosophical-political basis of this preparation [for the summit] were the decisions of the XXVII Congress of our Party *and the programme for a nuclear-free world set out on 15 January 1986* [emphasis added]. Only yesterday," he added, "this seemed to many to be utopia. Today it is becoming a fact."[80]

The essence of the 1986 statement is that it lays out a timetable and it links disarmament and development. It does this through two exactly matched sets of three rolling five-year plans to the year 2000. The first of the multistaged disarmament plans coincides with the 12th Five-Year Plan, the last with the 14th, in which the USSR is to attain a qualitative transformation of its economy by virtue of incorporation of scientific-technological advance in the previous 15 years. The explanation for the close coupling of the plans, Franks suggests, is self-evident: the disarmament plan releases resources at a quickening pace matched to the ability of the reviving economy to take them up productively.

Why so precise a target as the year 2000? In July 1986, during the period of currency of the unilateral Soviet moratorium on nuclear testing, Gorbachev met a group of scientists and strategists invited to Moscow to discuss prospects for a comprehensive test ban. He explained to them that he had been much criticized for utopianism in hazarding so precise a hostage to fortune as the target of the year 2000. But, he said, it was unavoidable for symbolic reasons; 2000 will be the turn not only of the century, but of the millennium. It is, he said, vital for mankind to have set down clear and visible markers by that moment to show that it has got a decisive grip upon the self-created problems of the second half of the twentieth century. We should not enter the next millennium carrying that baggage. He might have added that at the end of the millennium, he will have been in power for 15 years and be approaching the age of 70.

His timetable has already slipped, badly on the economic front, and there is no guarantee of his safe passage; but it should be remembered that at least an attempt has been made to effect a harmonized coupling of the dependent socioeconomic, military-technical, diplomatic, and political levels to a statement of long-term vision. Aspin's instinct is correct to think that the West needs similarly detailed and hard-headed planning, not just warm sentiments about the long march of freedom.

In the January 15, 1986 speech, Gorbachev also opened a theme that has become more pronounced over time. The global problems of the coming decades are beyond the competence of single states, or of an international system built on states' rivalry, to solve. The emergence of global interdependence as a theme in Soviet ideology predates Gorbachev's accession to power. Academicians at IMEMO (The Institute of World Economy and International Relations) were working on these questions in the mid-1970s. But the issue has been much developed under Gorbachev's patronage. Anatoly Dobrynin's appointment to head the International Department of the Central Committee in March 1986 gave a major voice to the globalist theme (called the "all-human problems" in the Russian literature). It was argued that under the threat of nuclear catastrophe, the old orthodoxy that a strengthening "correlation of forces" would bring the triumph of socialism was undermined. "Peace" gained more status as a social value and a more optimistic view was taken of the attainability of a stable peace with the capitalists. Greater weight came to be attached to the nonmilitary aspects of the "correlation of forces."[81]

What agencies, however imperfect, existed that might be adapted to the scale of the new task? Only one—the United Nations. But it had been crippled for many years by Soviet self-serving. Many unanswered questions remain about the Soviet role in Waldheim's career. It was following the new assessment in January 1986 that the Soviet attitude changed. Arrears were paid, the Security Council's potentials were praised, international law became a prominent theme in speeches, Soviet diplomacy through the UN became markedly more energetic. It was therefore entirely logical—and with hindsight predictable—that Gorbachev should use the UN podium to announce the conventional forces cuts and that he should embed that announcement within a speech otherwise entirely preoccupied with a wide-ranging and rather philosophical disquisition upon the powerlessness of single states and the obsolescence of military might in the attainment of influence. "A new way of thinking is not an improvisation, nor a mental exercise. It is a result of serious reflections on the realities of today's world," writes Gorbachev. Describing the source of this surprising empiricism, he adds poignantly ". . . it can be said that we have conceived the new mentality through suffering," without further elaborating its cause.[82]

CONCLUSION

The interloper stands free of shackles. The best of NOD conforms to the paradoxical logic of strategy better than any other military option currently on offer. These ideas offer a sound basis for a unilateral restructuring of NATO under the triple criteria proposed here. But it is most unlikely that that will be done. Whatever the formal logic of the case, it breaks too many emotional tendons in the West to adopt it.

NOD has not and will not transform the strategic outlook of the West directly for another reason also. The proponents remain, largely, trapped in the lower scales. Neither they nor the Western Official Mind have the capacity to devise a rigorous and integrated means of linking the scales of strategy to that of vision. But others have. Taken á la carte and blended with other sources of military thinking, the Soviet General Staff appears to be in the lead in exploiting the potentials of NOD to describe "reasonable sufficiency" in conformity with the new doctrinal guidelines set in May 1987. This is ironic, given that the ideas are Western and have only very recently entered the Soviet debate with any force.

Yet if the study of Western NOD proposals by Soviet civilian analysts has materially influenced the Soviet General Staff in ways that have helped it to agree to the deep cuts in conventional forces that Gorbachev desires for other, higher order reasons to do with his vision of a new Europe, then— double irony—the Noddists *do* have a chance and responsibility to play a role on the Western side.

Their ideas are indirectly reshaping the strategic level of Western thinking, via the classic, once-dreaded "indirect approach"—recycled in this way from the East. The effects of the dramatic changes in WTO doctrine are now beginning to be turned into decommissioned tanks. The more that this happens, the more seriously the Western Official Mind attempts to compute the NATO triangle of problems; and the more it fails at angles 1 and 2, the more interestedly it looks to angle 3.

Angle 3—preferably but not necessarily taken as a set menu—*can* compute angle 1. It cannot offer a strategy for victory on the European battlefield in the 1990s, but neither can anyone else. It can offer a way in which to travel with hope and confidence into the next, crucial phase of the CFE process in Vienna—the most crucial diplomatic negotiation of this generation for the future of Europe.

This is a precise task in time and space and purpose on the stage. In the great ballet of Europe Reborn, now playing, the role of NOD is to choreograph the difficult bits of the CFE *pas de deux* for the principal ballerina and her consort, the Two Europes themselves, East and West (who are linked romantically off-stage by persistent rumors in the popular press, most recently in Poland). NOD is not the prima ballerina assoluta, as some of its proponents once thought. Nor is the dance that NOD can choreograph the ballerina's main piece. She and her partner will dance in the next act the splendid vision of a new Europe, to grander music scored by other, nonmilitary hands. But they cannot do it with broken bones. Getting them safely through the present dance is all that is asked of NOD. *Ce n'est pas la guerre—mais c'est magnifique!*

Notes to Chapter 4

1. There are two specialized journals that conveniently embrace the range of debate about NOD. One is *Non-Offensive Defense: International Research Newsletter*, edited by B. M. Moller and published by the Centre of Peace & Conflict Research, University of Copenhagen. The other is *Defense & Disarmament Alternatives*, edited by K. White and published by the Institute for Defense & Disarmament Studies, Brookline, MA.

2. Defensive Defense; Nonoffensive Defense; Reactive Defense; Nonprovocative Defense; Structural Incapacity for Attack (*Strukturelle Nichtangriffsfähigkeit*, as we should all really learn to call it, being the leading usage in the FRG); Area Defense; Static Defense; Techno-Defense; Territorial Defense; Militia Defense; Nonnuclear Defense; Autonomous Defense; Transarmament. See J. Dean, "Alternative Defense: Answer to NATO's Central Front Problems?" *International Affairs* (Winter 1987-88), pp. 61–82. One should add the nice pun "Just Defense" to the list for completeness.

3. For the initial explanation in the West of the 1966 "decision point," subsequently broadly confirmed by other scholars, see M. MccGwire, *Military Objectives in Soviet Foreign Policy* (Washington DC, Brookings Institution, 1987).

4. H. Tromp, "Syllabus Errorum: Security Policy in the Nuclear Age," in *The Study of War & Peace: Essays from the 25th Anniversary Conference* (The Polemological Institute, University of Groningen, 1989). An earlier version of his paper was presented at 51st Pugwash Symposium: *Political Conditions for Peace & Security*, Bochum, FRG, April 21–24, 1988. An abbreviated form is available in the Précis of Lectures, Senior Officers' Conference 1988, G. Prins (ed), University of Cambridge Board of Extra-Mural Studies.

5. The classic accounts of this are in J. K. Galbraith, *The Great Crash, 1929*, rev. ed. (New York: Penguin, 1975); C. Kindleberger, *The World in Depression* (London: The Penguin Press, 1973).

6. "The United States today is headed for a financial and economic crisis. What appeared to be only a possibility five or six years ago became a probability more recently and has now become a virtual certainty. The only real questions are when and how." Felix Rohatyn, "On the Brink," *The New York Review of Books*, June 11, 1987, p. 3. In fairness, and parenthetically, one should add that John Kenneth Galbraith also warned that the Crash was imminent. He did so while at his summer home in Vermont in 1987. Unfortunately, his prediction only made headlines in the *Brattleboro Examiner*. (I am grateful to him for a sight of the article in question.)

7. The manner of Gorbachev's coming to power was, in fact, Stalin's last posthumous act in Soviet history: The "*perestroikists*" came to power as an age cohort over a five-year period because the cohort above them retired and died as a group. This was scarcely surprising, for as a group, Stalin's young tigers, later members

of Brezhnev's immediate postwar "Dnepropetrovsk mafia," had replaced the victims of the last and most debilitating of Stalin's purges of the Soviet leadership. Bullets in the neck in the late 1930s tidied up the subsequent cycles of succession.

8. On the supply side of Monetarist Beach, they found their old friends the Brits, who had also decided to take some time off from the wearisome business of having lost an empire and not yet found a role. Mrs. Thatcher wasted their windfall inheritance in the form of North Sea oil and with it, the chance for sensible economic reconstruction and a proper role in the new Europe, to engage in some reassuringly British pursuits on her Magical Mystery Tour. The money was used to pay for the sharpest depression of the century, which created the high unemployment essential for some splendid class warfare. There was lots of union bashing, much rearrangement of the furniture (called "privatization"); there was even a nice little imperial war. As a T shirt worn by some sailors in the South Atlantic accurately said: "The Empire strikes back"—with a vengeance, and in more ways than one. But the trip is nearly over and the British, much as it irks them, now do not really count. Thatcher has marooned them in weak eccentricity. So they go into a footnote.

9. The terms are from L. J. Dumas, *The Overburdened Economy* (Berkeley: University of California Press, 1986). It has long been thought that Marshall Aid was axiomatic to postwar European reconstruction. A. S. Milward *The Reconstruction of Western Europe, 1945–51*, California, 1984 has shown that the depth of infrastructural damage was not as great as it appeared and that human, not material capital was the vital prerequisite for swift recovery.

10. P. Odell, *Oil and World Power*, 7th ed. (New York: Penguin 1983), pp. 62–65; M. Walker, *The Waking Giant: The Soviet Union under Gorbachev* (Michael Joseph, 1986) tells the fertilizer story.

11. Further discussion of this will be found in E. Luard, *The Blunted Sword: The Erosion of Military Power in Modern World Politics*, I. B. Tauris, 1988 and in A. Giddens, *The Consequences of Modernity*, Polity Press, forthcoming 1990.

12. For a similar conclusion, argued from the main premiss of Soviet changes, see M. MccGwire, "The New Challenge of Europe," in A. Clesse and T. C. Schelling (eds): *The Western Community & the Gorbachev Challenge* (Baden-Baden: Nomos Verlagsgesellschaft, 1989).

13. "A Comprehensive Concept of Arms Control & Disarmament, adopted by Heads of State and Government at the Meeting of the North Atlantic Council in Brussels on 29th and 30th May 1989," NATO Information Service, 61, p. 14.

14. A convenient summary is found in J. Boutwell et al., *Countdown on Conventional Forces in Europe: A Briefing Book*. American Academy of Arts & Sciences/Ploughshares Fund, November 1988, pp. 15–20.

15. J. Dean, "The Vienna Force Reduction Talks—Resolved and Unresolved Issues," paper to the 39th Pugwash Conference on Science and World Affairs, Cambridge (MA) July 23–28, 1989, p. 2. A detailed exposition of the steps in the negotiation May—July 1989 is given in the invaluable and timely *Cutting Conventional Forces An analysis of the Official Mandate, Statistics, and Proposals in the NATO-WTO Talks on Reducing Conventional Forces in Europe* (Brookline, MA: Institute for Defense & Disarmament Studies, 1989), pp. 15–36.

16. M. Gorbachev, Address to the 43rd session of the UN General Assembly, New York, December 7, 1988, Novosti Press Agency, pp. 20–21.

17. "Comprehensive Concept," ¶ 11, p. 3.

18. Proposal of 13 July 1989, Ch 4., p. 5.

19. Major General A. J. G. Pollard, "Future Defense of the Central Region," *Conference on Britain and the Central Region* London: British Army Staff College, in association with King's College, London, at the University of London, 1989).

20. A partial survey of the main actors in the debate was provided by Col H. D. Lemke, "Modes of Defense: Aspects of the West German debate," *Conference on Britain & the Central Region,* July 18 / 19, 1989.

21. Pollard discriminates semantically between the terms "mobile" and "maneuver." He locates "mobile defense" as a tactical opportunity within the overall strategic concept of "maneuver warfare."

22. Pollard, p. 4.

23. Pollard, p. 2; General Sir Martin Farndale, "Command and Control of the Joint Army Group / Tactical Airforce Battle at the Operational Level," *Conference on Britain & the Central Region,* July 18 / 19, 1989.

24. Pollard, p. 14.

25. ". . . it is his leaders who must believe in the threat . . . and this makes security obtained by suasion inherently less reliable than defensive denial would be." E. Luttwak, *Strategy: The Logic of War and Peace* (Cambridge: Belknap / Harvard University Press, 1987), p. 122 *passim*. With French strategic thinkers, Luttwak rightly abjures the sloppy usage of "deterrence," preferring instead "suasion," since *suadere* includes and exceeds the scope of *terrere* and is therefore a word more realistically descriptive of the spectrum of interstate signaling. An extended critique of Luttwak's thinking is given in G. Prins, "Perverse Paradoxes in the Application of the Paradoxical Logic of Strategy," *Millennium: Journal of International Studies,* 17,3 (1988): 539–551.

26. A brief history of three phases in Alternative Defense thinking in the FRG since 1952 and the creation of *Amt Blank,* illustrating this point, is found in D. Stevenson, "A Conflict of Priorities: The West German SPD and Security Policy," *Peace Research report No 22,* School of Peace Studies, University of Bradford, March 1989. pp. 42–53.

27. Bundeswehr / Emdit poll, December 1988.

28. Pollard, pp. 7–9 and Figures 5 and 6.

29. U.S. Army Field Manual FM 100 / 5, 1982.

30. Air Chief Marshal Sir Anthony Skingsley, "Interdiction and Follow-On Forces Attack," 1245.2 / ACPOLT / 74 / 89 NATO unclassified, June 22, 1989.

31. P. Sharfman (The MITRE Corp), "The Future of FOFA," *Conference on Britain & the Central Region,* July 18 / 19, 1989. pp. 1–10.

32. Skingsley, p. 9.

33. P. Bracken, "Deep attack and European security," *Mss.* (November 1987). The consequence for declaratory policy clarified in discussion.

34. Sharfman, p. 7.

35. Bracken, pp. 38–67. A new, integrated battlefield C2 system—BICES—with 1OC 1992 is presently in development. Its proponents believe that it will be a dynamic, common data base that will reduce the vulnerability of nodal points. This remains to be seen, but past experience is certainly against it.

36. McGeorge Bundy, George Kennan, Robert McNamara, and Gerard Smith, "Nuclear weapons and the Atlantic Alliance," *Foreign Affairs,* vol 60, (Spring 1982); Field Marshal Lord Carver, *A policy for peace,* (London: Faber & Faber, 1982), pp. 102–3. Major statements signaled the shift in the West German position: Disavowing the battlefield nuclear rung, General Altenburg stated ". . . whoever thinks that way and hence assigns nuclear weapons the role of a better sort of artillery, has not understood their purpose and their role within our strategy." Speech to the Atlantic Treaty Association, September 1986. The most perceptive discussion of these changes in the Federal Republic is to be found in M. Lucas, "The United States and Post-INF Europe," *World Policy,* (Spring 1988), pp. 183–233.

37. C. M. Kelleher, "NATO Nuclear Operations" in A. Carter, J. Steinbrunner, and C. Zraket (eds): *Managing Nuclear Operations* (Washington, DC: Brookings Institution, 1987), pp. 445–69; D. Ball, *Controlling theatre nuclear war.* Working paper No 138, The Strategic and Defense Studies Centre, Australian National University, 1987.

38. Ball, p. 7, p. 31.

39. Kelleher in Carter, Steinbrunner, and Zraket, p. 469.

40. Luttwak, p. 174.

41. ". . . FOFA is simply an evolution of interdiction that focuses more clearly on the critical elements of the enemy's attack plans. It is thus very much part of a defensive campaign, although it involves operations deep into enemy territory." Skingsley, p. 7; Pollard, p. 11.

42. A. A. C. von Müller, "Conventional Stability & Arms Control. Outlines of an Integrated Security and Defense Policy," mss. Starnberg 1987; His ideas are most recently published in A. von Bülow, H. Funk, and A. A. C. von Müller, *Sicherheit für Europa* (Koblenz: Bernard & Graefe Verlag, 1988).

43. L. Unterseher, "Spider & Web: The Case for a Pragmatic Defense alternative," (Bonn: SAS, 1988); L. Unterseher, Diagram, "The SAS Concept: Structural Change for the Nineties," presented at International Studies Association Conference, London, March 28–April 1, 1989; SAS, *Vertrauenbildende Verteidigung: Reform Deutscher Sicheheitspolitik,* (Gerlingen, 1989). (This book is reviewed in *NOD–International Research Newsletter,* May 12, 1989, p. 53.) For a comprehensive review of current German proposals from a somewhat different perspective, K. Voigt, "Konventionelle Stabilisierung und strukturelle Nichtangriffsfähigkeit: Ein systematischer Vergleich verschiedener Konzepte," Aus Politik und Zeitgeschichte, B 18/88, April 29, 1988, pp. 21–34. (I am grateful to E. Boeker for alerting me to this essay.)

44. The new mines are to be seen in action in a BBC "Horizon" program made by Peter Ceresole. "Smart Weapons," January 26, 1989, no. 50/LSF/A405.

45. R. K. Hüber and H. W. Hofman, "Gradual Defensivity: An Approach to a Stable Conventional Force Equilibrium in Europe? *Operations Research* (1984), pp. 197–211; H. W. Hofman, R. K. Hüber, and K. Steiger, "On Reactive Defense Options—A Comparative Systems Analysis of Alternatives for the Initial Defense Against the First Strategic Echelon of the Warsaw Pact in Central Europe," in R. K. Hüber (ed): *Modelling & Analysis of Conventional Defense in Europe* (New York: Plenum, 1986), pp. 97–140.

46. H. W. Hoffman and R. K. Hüber, "On the Role of New Technologies for Conventional Stability in Europe," mss. Universität der Bundeswehr, München, April 1988.

47. R. K. Hüber and H. W. Hoffman, "Some Thoughts on Force Modeling for Arms Control," mss. Universität der Bundeswehr, München, 1988.

48. R. K. Hüber, "On Strategic Stability in Europe without Nuclear Weapons" in R. Avenhaus, R. K. Hüber, and J. D. Kettelle (ed): *Modelling & Analysis in Arms Control*, NATO Advanced Science Institutes Series F, Vol 26 (New York: Springer-Verlag, 1986), pp. 215–39; R. K. Hüber, "On Structural Prerequisites to Strategic Stability in Europe without Nuclear Weapons: Conclusions from the Analysis of a Model of Conventional Conflict," *Systems Research* 5,iii (1988); 1–13.

49. Interview with Philip A. Karber by Benjamin F. Schemmer, March 24, 1987, published in *Armed Forces Journal* (May & June 1987). See esp. map and discussion of it at p. 116, June 1987 issue.

50. General Sir Hugh Beach, "On Improving NATO Strategy," in A. J. Pierre (ed): *The Conventional defense of Europe: New technologies and new strategies*, Europe / America series No 5, Council on Foreign Relations, (New York: New York University Press, 1986), p. 179.

51. Luttwak, p. 236.

52. Ibid., p. 122.

53. A. Wohlstetter and F. Iklé (eds), *The Commission on Integrated Long-Term Strategy, Discriminate Deterrence* (Washington DC: USGPO, 1988), pp. 6–11; P. Kennedy, *The Rise and Fall of the Great Powers: Economic Change and Military Conflict, 1500–2000* (London: Unwin, 1987) (the idea of "imperial overstretch" is, of course, original to Edward Gibbon); John Palmer, *Europe without America? The Crisis in Atlantic Relations* (New York: Oxford University Press, 1988); D. P. Calleo, *Beyond American hegemony: The future of the Western Alliance* (New York: Basic Books, 1987), p. 168.

54. M. van Creveld, *Supplying War: From Wallenstein to Patton* (Cambridge: Cambridge University Press, 1977).

55. M. van Creveld, "Logistics since 1945: From Complexity to Paralysis?" Conference on Britain and the Central Region, July 18 / 19, 1989.

56. Major General J. Thompson, "Command and Control (final version)," Conference on Britain and the Central Region, July 18 / 19, 1989, pp. 12, 13.

57. van Creveld, "Logistics since 1945," pp. 4–5.

58. van Creveld, "Logistics since 1945," p. 11. In fairness it must be stated that this point of view is energetically rejected by officers with logistics responsibilities.

59. In answer to the author in open session at the ISA conference in London, March 1989.

60. An exception is G. Prins, "Naval Forces and Defensive Defense," *Bulletin of Peace Proposals*, 1990.

61. I am indebted for the following example and much of the analysis to a mss. version of parts of M. MccGwire, *Perestroika and Soviet National Security* (Washington DC, Brookings Institution, 1990).

62. *The Times*, December 17, 1987, p. 16. I am indebted to Peter Frank for this reference.

63. M. Gorbachev, *Perestroika: New thinking for our country and the world* (New York: Harper & Row, 1987), p. 195.

64. For example, S. Karaganov, "The Common European Home: The Military Angle," *International Affairs*, August 8, 1988; A. Arbatov, "Parity and Reasonable Sufficiency," *International Affairs*, October 10, 1988.

65. Correspondence between Gorbachev and Mssrs Boserup, von Hippel, von Müller, and Neild, December 1987.

66. IMEMO, *Disarmament & Security. 1987 Yearbook* (Moscow: Novosti, 1988). Alexei Arbatov writes the chapter on doctrine, in which he concedes that the Soviet "offensive defense" will ". . . inevitably be perceived by other countries as a sign of aggressive intentions" (p. 215). Mssrs Amirov, Kishilov, Makarevsky, and Usachev in their chapter on reducing confrontation conclude that adoption of NOD is the best way to go (p. 390).

67. S. Shenfield, *The Nuclear Predicament: Explorations in Soviet Ideology*, Chatham House Papers No 37 (London: Routledge & Keegan Paul, 1987). This book is a brilliant and timely piece of detective work, explaining the intellectual and bureaucratic/political evolution of the "new thinking" of *perestroika*.

68. MccGwire, *Military Objectives* . . . "A Potential Decision Period: 1983–85," pp. 307–34.

69. P. A. Petersen, "Soviet Military Thinking in an Age of New Thinking on International Security," *Maritime Conference 1989: Response to Change in the 1990s*, (Greenwich: Royal Naval College, 1989).

70. The following is indebted to John Erickson in *Précis of Lectures: Cambridge Senior Officers' Conference 1988*, G. Prins (ed) (Cambridge: Extra-Mural Board, University of Cambridge).

71. C. N. Donnelly, "Analysis of Soviet Military Thinking in 1989," Conference on Britain and the Central Region, July 18/19, 1989.

72. S. Larrabee, "Gorbachev and the Soviet Military," *Foreign Affairs*, 66, 5 (Summer 1988).

73. Gorbachev, *Perestroika* . . ., pp. 49–55, describes *perestroika* as an extension and development of the 1917 Revolution.

74. A point stressed by deputy chief of the General Staff, Colonel General M. Gareyev, in speeches on a West European tour in the autumn of 1988 and in "The Armed Forces in the Conditions of Glasnost," *Argumenty i Fakty*, 39, September 24–30, 1988, *cit* Petersen, "Soviet Military Thinking in an Age of New Thinking."

75. Shenfield, *The Nuclear Predicament,* substantiates this line of analysis.

76. R. Allason, seminar on developments in Soviet military thinking, King's College Research Centre, Cambridge, June 20, 1989.

77. Gorbachev, address to the 43rd session of the UN General Assembly, New York, pp. 20–21.

78. Hon L. Aspin (chairman, Committee on Armed Services, U.S. House of Representatives), "Meeting Gorbachev's Challenge," Brussels, December 15, 1988.

79. P. Frank, "Address on the Reasons for *perestroika," Maritime Conference 1989: Response to Change in the 1990s* (Greenwich: Royal Naval College, 1989).

80. *Pravda,* December 15, 1987, *cit* Frank, "Address on the Reasons for *perestroika,"* p. 11.

81. This paragraph summarizes arguments from Shenfield, *The Nuclear Predicament.*

82. Gorbachev, *Perestroika: New Thinking for Our Country and the World,* p. 145.

Part III

Defensive Strategies in Regions of Conflict: Northeast and Southeast Asia

Chapter 5

Nonprovocative Defense Strategy

J. Soedjati Djiwandono

This chapter offers some thoughts on nonprovocative or nonoffensive defense strategy, its credibility, and possible application to South and Southeast Asia, particularly in areas of actual or potential conflicts. Such an examination cannot be explored in all its aspects due to space limitations and because a common understanding or agreement seems yet to be reached on what are the proper components of nonprovocative or nonoffensive defense strategy.

The idea of a nonprovocative defense or one under a similar label has grown out of the Western European context, situation, and experience, particularly with the advent of new nonnuclear weapons. However, although perhaps not using the same terminology, some countries in South and Southeast Asia have tried to develop and pursue a defense policy and adopt a defense posture at one time or another with basically the same principles underlying the nonprovocative approach, particularly in various forms of territorial defense. In Southeast Asia, for example, Indonesia has pursued what is termed an "all-people's defense and security system"; Malaysia has developed the principle of doctrine of "comprehensive security," which reminds us of Japan's security principle under the same label. That such a doctrine is basically nonprovocative, nonoffensive, or manifestly defensive and serving primarily as a deterrent against aggression is reflected in the remark made by the former deputy prime minister of Malaysia, Musa Hitam, who referred to the aim of the doctrine as being to make the country "like a lump of steel—possible to swallow but impossible to digest."[1] Likewise, having decided to create a citizen army, Prime Minister Lee Kuan Yew of Singapore compared his country to a poisonous shrimp: larger fish that preyed on it would die.[2] However, such factors as geographical and geopolitical situations, historical backgrounds, and certain aspects of defense policy pursued by some countries in South and Southeast Asia, in spite of the adoption of basically and originally a nonprovocative or nonoffensive defense, seem so incompatible with that very defensive strategy as to reduce, to say the least, its credibility. As a consequence, the applicability of nonprovocative defense to the majority of countries in the region seems to be in doubt. Therefore, an alternative arrangement for the security of these countries, for each region collectively more than individually, is suggested here without abandoning the nonprovocative, nonoffensive principle. Indeed, such an arrangement would even more strongly emphasize its nonprovocative and inoffensive nature.

IN SEARCH OF A PARADIGM

It may be presumed that nonprovocative defense is an appropriate policy for small and medium-size states in particular. Thus on this account alone, in South Asia nonprovocative defense would not be appropriate for India, definitely a big power. In theory at least, India may face the threat of an invasion by the Peoples Republic of China (PRC) or at least get involved in an armed conflict with it over a border dispute such as in 1962. For that reason, India may develop a nonprovocative defense posture as a deterrent against such an invasion. But because India is such a large country compared to its neighbors, its defense posture, regardless of intentions, would hardly be perceived as nonprovocative and inoffensive by such states as Pakistan, Sri Lanka, Bhutan, and Nepal. And its civil resistance during its struggle for independence against British colonial rule notwithstanding, which was certainly one form of nonprovocative defense, India's invasion of Bengal in 1971 (which ultimately gave birth to Bangladesh), regardless of its moral justification, would probably exclude it from those states pursuing a nonprovocative and inoffensive defense policy.

The same is true for Vietnam in Southeast Asia, which is not really a big state, although the largest among the Indochinese countries. However manifestly defensive and inoffensive its previous struggle against the Japanese, the French, and the Americans might have been, its defense strategy can hardly be any longer regarded as nonprovocative and inoffensive since its invasion and occupation of Kampuchea at the end of 1978, again regardless of moral justification. This is certainly true at least as far as Thailand's perception is concerned.

A similar case may be made with respect to Indonesia, the largest country in Southeast Asia in terms of territory, population, and natural resources. As referred to earlier, Indonesia's defense and security policy is one of territorial defense, designated as an all-people's defense and security system, which is one form of nonprovocative and inoffensive defense, particularly developed on the basis of its revolutionary experience during the struggle for independence against Dutch colonial rule. Yet because of its size, whether or not its defense policy and posture should be regarded as nonprovocative and inoffensive would be determined primarily by the perception of its smaller neighbors, that is to say, whether or not its defense policy and posture provide adequate confidence and assurance on the part of its neighbors. Apart from the question of size, which is an important factor determining the nature of relations between states, any possible lingering suspicion of Indonesia's future intentions and a perception of Indonesia as a potential threat on the part of its neighbors in Southeast Asia and beyond (such as Papua New Guinea and Australia) would be understandable against recent historical backgrounds. Under President Sukarno in the midcentury, Indonesia pursued a policy of confrontation against Malaysia. Its struggle for the recovery of West Irian—now the province of Irian Jaya—is not always correctly appreciated, and a number of countries still find it hard to acquiesce to the integration of East Timor with Indonesia in 1976. They tend to see it as pure annexation of the territory by Indonesia, thereby betraying its aggressive, offensive, and expansionist designs. Thus on those accounts, although probably possessing certain other qualifications and despite its adoption of territorial defense, Indonesia is not easily included in the category of states pursuing nonprovocative and inoffensive defense strategy.

It seems to go without saying that nonprovocative and inoffensive defense strategy should exclude the possession or deployment of nuclear weapons, for even nuclear weapons of a "tactical" nature would possess an offensive capability. Against what is supposed to be a superior conventional force, even tactical nuclear weapons would serve as a deterrent through punishment rather than through denial, and thus have offensive and escalation capabilities. And given the suspicion that in South Asia, India and perhaps Pakistan may very well "go nuclear" would render their defense strategy provocative and offensive. It has been said about territorial defense, which may be regarded as a form of nonprovocative defense, that it is a possible form of "unconventional deterrent" and a way of "escaping or reducing the danger of defending one's country by apocalyptic threats."[3]

Alliance relationship seems incompatible with nonprovocative defense, especially if—as normally understood—alliance involves guarantee by a superpower with strategic and offensive capabilities. Such an alliance would entail intervention and retaliation, which certainly render the label nonprovocative or inoffensive pointless and meaningless. Thus though small and nonnuclear, Thailand and the Philippines in Southeast Asia would not likely qualify for being nonprovocative and inoffensive in their defense policies and postures on account of their alliance with the United States. And apart from the reasons already noted, the same is true with Vietnam because of its alliance with the Soviet Union, which had enabled it to invade and occupy Kampuchea.

Although it is not possible here to mention all the states in South and Southeast Asia in relation to nonprovocative defense, at least two more are worth noting—Afghanistan in South Asia and Kampuchea in Southeast Asia. Both have been involved in internal and to some extent external conflicts for the past decade. In each case, the general image seems to be purely or at least primarily a conflict between guerilla resistance and an occupying power, the Soviet Union in Afghanistan and Vietnam in Kampuchea. Some appreciation has been given to the strength of the resistance forces in both cases and the difficulties they have created for the occupying forces as well as for the regimes they have created or helped sustain.[4] In fact, such has been the image that the central issues debated over the years have been the Soviet invasion and occupation of Afghanistan and the Vietnamese invasion and occupation of Kampuchea. This is not a complete and accurate picture, however, for the Soviet withdrawal from Afghanistan in early 1988 has not resulted in the cessation of the conflict in that country and the Vietnamese withdrawal from Kampuchea has not been followed by an immediate end of the conflict either. One of the curious things about the Afghan and Kampuchean conflicts is that although the Soviet-backed regime in Kabul is widely recognized by the international community including the United States, international recognition has been accorded to the so-called CGDK, the resistance forces against Vietnamese occupation as well as against the Vietnamese-backed Heng Samrin regime in Phnom Penh. Wide international support and sympathy have been given to the resistance forces in both Afghanistan and Kampuchea.

Whatever the case, under the circumstances both Afghanistan and Kampuchea being divided nations, it would seem of little relevance to discuss the question of nonprovocative defense with regard to either one. In fact, because the resistance forces in both cases are externally based, although all the parties to the conflicts are externally backed, it would even seem to make more sense to talk about defense with reference to the regimes in power in the face of external attacks.

Another divided nation, though not to the same extent of Afghanistan or Kampuchea, is Sri Lanka. Any discussion of nonprovocative defense would also seem irrelevant under the present conflict conditions between the Tamil population and the central government. Under such circumstances, an invasion or intervention by a foreign power, for example, India, might not be necessarily unwelcome, particularly by some part of the population. A not entirely different case may be made for Pakistan. By hindsight, events in 1971 that led to the creation of Bangladesh might raise questions as to whether it had been realistic to talk about Pakistan as a nation, for which the question of nonprovocative defense would have been of little relevance as well.

Moreover, when Z. A. Bhutto pushed the idea of a "nation at arms," a form of nonprovocative defense, his aim was perhaps in part internal, which was to curb the military leadership from playing too great a role in the country's domestic politics, an aim that Prime Minister Lee Kuan Yew of Singapore seems to have shared at least partially.[5]

Thus it seems clear that nonprovocative or nonoffensive defense strategy would lack credibility in South and Southeast Asia on account of varying perceptions based on such factors as historical backgrounds, and geographical and geopolitical conditions. This would raise doubts about its applicability to the two regions, particularly with reference to its possible practice by the individual countries.

REGIONAL ALTERNATIVES

It seems clear that a key issue in nonprovocative or nonoffensive defense strategy is one of perception. That is to say, a state's defensive strategy is nonprovocative and inoffensive, quite apart from the question of force structure, weapons, and delivery systems, primarily to the extent that other neighboring states, potential adversaries or otherwise, have the confidence and assurance that the defensive strategy is indeed manifestly defensive, nonprovocative, and nonoffensive. This is true particularly as regards the countries of South and Southeast Asia, whose perceptions of threat to their security, not only internal and beyond the regions, but also, especially in the context of nonprovocative defense, from one another, are shaped by historical experience, and their locations and politics. The most important question, therefore, is one of confidence building.

As far as Southeast Asia is concerned, the establishment of the Association of Southeast Asian Nations (ASEAN) regional cooperation—Burma and the Indochinese countries still excluded—seems to be the best and so far the most successful means of confidence building. Indeed, concealed behind its rhetoric on economic and socio-cultural cooperation, the establishment of ASEAN was motivated primarily by political and security considerations, although for some understandable reasons these aspects of ASEAN regionalism were never explicitly stated.[6] The founding Bangkok Declaration of August 1967 declares that the member states of ASEAN "are determined to ensure their stability and security from external interference in any form or manifestation in order to preserve their national identities in accordance with the ideals and aspirations of their peoples." In fact, the threat of external interference has been a major preoccupation of the ASEAN states as reflected in its repetition and reaffirmation in subsequent ASEAN documents, particularly on the principle of ZOPFAN (ASEAN's proposal on the establishment of Southeast Asia as a zone of peace, freedom, and neutrality), the Declaration of ASEAN Concord, and the Treaty of Amity and Cooperation in Southeast Asia, both signed at the first ASEAN Summit in Bali in 1976. But what have given room for external interference in Southeast Asia in the history of the region since the end of World War II have been primarily domestic and interstate conflicts.

The establishment of ASEAN, therefore, should be seen against the background of a situation of conflict in the region immediately before, namely, Indonesia's confrontation against Malaysia. It was an intraregional conflict that had invited the increasing presence and role of great powers in the region. It had thereby provided an opportunity for great power intervention, either directly, as in the case of Great Britain, which came to the aid of Malaysia, or less directly as in the case of the Soviet Union, then supplanted by the PRC, which came to the aid of Indonesia. Indeed, neither the Soviet Union nor the PRC was directly involved in the Indonesia-Malaysia conflict. Yet it was certainly the conflict situation, not only in terms of the intraregional conflict but also in terms of tension among the socio-political forces in Indonesia not unrelated to the former, that had made room for the alleged Chinese involvement in the abortive coup attempt by the Indonesian Communist party in 1965.

Thus the establishment of ASEAN could not be separated from the end of Indonesia's policy of confrontation against Malaysia. At least as far as Indonesia is concerned, ASEAN may be said to serve as evidence and assurance of its good faith and determination to bring an end to its confrontation policy and to pursue, instead, a good-neighbor policy. Indonesia's commitment to this new peaceful policy is reflected to the symbolic significance of the fact that an exchange of diplomatic representation between Malaysia and Indonesia was effected after, not before, the establishment of ASEAN itself.

It may be argued that as for other member states of ASEAN, especially Malaysia, Singapore, and Brunei, which had been targets of its policy of confrontation, Indonesia's membership would reduce the possibility of threat to their security posed by their giant neighbor. In other words, within ASEAN, Indonesia would be domesticated as it were, and rendered less harmful, dangerous, and menacing than outside of it. The establishment of ASEAN was to serve as a guarantee against the recurrence of intraregional conflicts in the future.

To the extent that ASEAN has been successful in avoiding any further use or threat of force or violence in the settlement of intraregional conflicts, its regionalism has been a success. It explains, at least in part, the absence or lack of further external interference. And in that way ASEAN can ensure the security of its member countries.

To attain that primary goal, various forms of cooperation in various fields are to be carried out within the framework of ASEAN. As stated in the

Bangkok Declaration, the aim to "accelerate economic growth, social progress, and cultural development in the region through joint endeavors in the spirit of equality and partnership" was to "strengthen the foundation for a prosperous and peaceful community of Southeast Asian nations." In other words, the forms of cooperation in various fields as spelled out in the declaration are aimed at supporting the final cause of ASEAN, which is basically political in nature and security oriented.

It is true that in many instances political cooperation in the ASEAN framework has been more in the nature of reaction to external challenges than initiatives on the part of ASEAN, such as the Kampuchean conflict. But the development of common views and common stands in response to external challenges contains in itself an element of initiative and further encourages more initiatives. As a newly developing force, ASEAN may still be unable to make decisions or to take initiatives affecting international relations in a meaningful sense. Nevertheless, its reaction to external challenges may reinforce the strength and help advance the cause of ASEAN.[8]

First, external challenges have helped to promote ASEAN solidarity because of the development of common views and common stands. This is in part a realization of the program of action in the political field as adopted in the Declaration of ASEAN Concord for the "strengthening of political solidarity by promoting the harmonization of views, coordinating position and, where possible and desirable, taking common actions."

Second, such external challenges have encouraged the promotion of mutual consultations within ASEAN at various levels for the purpose of attaining those objectives. In other words, within ASEAN at least a trend seems to have developed toward the institutionalization of regular, intensive, and continuous interaction and intercommunication among its member countries.

Third, the trend toward the institutionalization of interaction and intercommunication as well as common interest in the promotion of solidarity among the member countries of ASEAN have helped to iron out differences and even to neutralize conflicts. Indeed, as stated in the Declaration of ASEAN Concord, "settlement of intraregional disputes by peaceful means as soon as possible" forms part of the program of action in the political field adopted by the Bali Summit, which also declares that the member states undertake to "rely exclusively on peaceful processes in the settlement of intraregional differences." In addition, the Treaty of Amity and Cooperation in Southeast Asia, which was also signed at the Bali Summit, constitutes, among other things, an attempt to create a mechanism for the peaceful settlement of intraregional disputes. It reaffirms the principles of "settlement of differences or disputes by peaceful means" and the "renunciation of the threat or use of force."

It is true that so far such a mechanism has never been utilized by any

Asean member state involved in a dispute such as Malaysia and the Philippines over such. Yet those factors have at least helped not only to neutralize intraregional conflicts but also to prevent the emergence of new conflicts. Of even greater importance, given the fact that the kind of mechanism created is as yet unable to settle intraregional conflicts peacefully as originally expected, such factors will help prevent or at least reduce the possibility of a resort to force or violence.

Thus it can be argued that a trend seems to have developed within ASEAN toward the establishment of some form of a "security community," in which expectations of the use of force for the settlement of conflicts among member nations are so remote and unthinkable that they are almost entirely eliminated. It is obvious that a security community is not a security or military pact, which was never the intention of ASEAN to create.

It does not mean, however, that there is no cooperation in the security field. Cooperation has been fostered on a bilateral basis in the framework of ASEAN. The primary reason for a lack in multilateral security cooperation is the absence of a common perception of external threat among the member states, which would serve as a glue to bind them together in a defense and security cooperation on a multilateral basis.[9] The only common perception of external threat among the member states is in the form of external interference. There is, indeed, a recognition of a common perception of internal threat as stated in the Declaration of ASEAN Concord, which says, inter alia, that "The stability of each member state and of the ASEAN region is an essential contribution to international peace and security. Each member state resolves to eliminate threats posed by subversion to its stability, thus strengthening national and ASEAN resilience." The declaration, therefore, only affirms the "continuation of cooperation on a non-ASEAN basis between the member states in security matters in accordance with their mutual needs and interests."

Security cooperation on such a non-ASEAN, thus basically bilateral, basis has taken various forms such as security arrangement of common borders, joint patrol, exchange of visits and personnel, education, research, and technical assistance, joint military exercises, and exchange of information and intelligence. These would greatly contribute to confidence building, which in turn would help reduce and over the long run may even eliminate perceptions of threat among the ASEAN member states. With mutual trust and confidence, whatever defense posture and policy each of the member states may adopt and pursue, especially in continuous mutual consultations and exchange of information, not only in security but also political matters, would less likely arouse suspicions on the part of one another as to their respective intentions. In other words, the degree of the nonprovocativeness and inoffensiveness of their defensive strategy would be increased.

To be sure, the ideals and principles underlying ASEAN regionalism have been applied only to its member states. But these ideals and principles have always been intended for Southeast Asia as a whole to include Burma and the Indochinese countries. ASEAN states certainly cannot claim a monopoly over what has conventionally be designated as Southeast Asia.

Without overlooking the real differences between South Asia and Southeast Asia in a number of respects, the same or similar ideals and principles underlying ASEAN regionalism may perhaps be applicable also to South Asia, where a similar arrangement in the form of SAARC has been developing. This is the sort of arrangement that may be not only a plausible alternative but preferably also a complement to the defensive strategies of the individual countries of the two areas, to give it a regional framework that would ensure their nonprovocative and inoffensive nature.

CONCLUSION

Such continuous efforts at confidence building as provided by regional cooperation will be all the more necessary in view of the fact that it is not always easy to draw a clear line between the defensive and the offensive nature of weapons, particularly in the light of rapid technological advancement in the new generation of conventional weapon systems. What purport to be defensive weapons have often been used for offensive purposes. In the Falkland Islands in 1982, for example, British forces used antitank missiles to attack fixed Argentine positions near Port Stanley, which was a tactically offensive use of such supposedly defensive weapons.[10] Or what basically is a defensive strategy may turn out to be offensive as well, as manifested by the Israeli invasion of Lebanon in 1982 or the Israeli bombing of what was thought to be an Iraqi nuclear plant near Baghdad.

Given the success, at least over the long run, of regional cooperation such as ASEAN in Southeast Asia and SAARC in in South Asia in developing a manifestly defensive strategy and posture, it would not imply that the regions or the countries therein would be made invulnerable either collectively or individually to external aggression in whatever form. But the strengthening of regionalism would certainly increase the costs of such an aggression. In that sense regionalism may function as a deterrent against such a threat.

Implicitly and indirectly, regionalism for the purpose of strengthening nonprovocative defense would result in disarmament and arms control. But whether nonprovocative defense in general should lead eventually to general and complete disarmament is open to serious question, especially because the idea of general and complete disarmament is itself of doubtful value in a world of sovereign and independent states that constitute what has been termed an anarchical society.

Finally, the full implications of nonprovocative defense would continue to merit serious study in the light of rapid developments in international relations. Would nonprovocative defense, for instance, continue to be relevant to claims to exclusive economic zones, which might require the backing of basically offensive capabilities? Or to disputes arising from territorial claims and counterclaims to island territories well beyond the traditional national boundaries of states making such claims and counterclaims, such as to the Spratleys and the Paracels in the South China Sea? These will be some of the challenges of the future.

Notes to Chapter 5

1. Quoted in Noordin Sopiee, "Malaysia's Doctrine of Comprehensive Security," *The Journal of Asiatic Studies*, XXVII, 2: 261.

2. Cited by Adam Roberts, *Nations in Arms: The Theory and Practice of Territorial Defence*, 2nd. rev. ed. (London: MacMillan for the International Institute for Strategic Studies, 1986), p. 33.

3. McGeorge Bundy in his Foreward to Roberts, *Nations in Arms*, p. 7.

4. *Ibid.*, pp. 280–82.

5. *Ibid.*, pp. 32–33.

6. I have dealt with these issues in a number of articles; see, e.g., "Indonesia's Relations with Other Southeast Asian Countries," in Jae Kyu Park and Melvin Gustov, (eds): *Southeast Asia in Transition: Regional & International Politics*, (Seoul, Korea: The Institute for Far Eastern Studies, Kyung Nam Unvarsity, 1977); "The Political and Security Aspects of ASEAN: Its Principal Achievements," *The Indonesian Quarterly*, Vol. XI, No. 3 (1983).

8. *Ibid.*

9. See J. Soedjati Djiwandono: "ASEAN's Security Concerns and Resources," in *Security of the Western Pacific Reexamined* (Taipei: The Peace Council, 1985), pp. 115–28.

10. Roberts, *Nations in Arms*, p. 290.

Chapter 6

International Relations in Asia

Tomohisa Sakanaka

International relations are changing rapidly, perhaps more so in Asia than in Europe. U.S.-Soviet progress in arms control promises a new era of superpower detente. After signing the INF Treaty, the governments of the United States and the USSR continue to negotiate a reduction of strategic nuclear forces and show signs of a real commitment to conventional arms reductions. Although it may be overly naive to assume that detente will automatically result from these discussions, the U.S.-Soviet dialogue is likely to continue and we can expect more positive developments in the future. Of equal importance perhaps in Asia in the rapprochement between the Soviet Union and China. The end to over 30 years of confrontation between these two major powers will have wide-ranging repercussions on the configuration of relationships in the region.

However, we should not be content simply to watch these changes unfold. Although these are positive developments, they do not completely erase the potential for instability in the Asian region. Japan and the United States should establish a consistent and long-term policy that will be able to meet the challenges that these changes present. The basic requirements of such a policy are: first, to further improve the current system of security cooperation to make it more adaptable to the changing environment, and, second, to eliminate mutual distrust and dissolve tensions between those states in confrontation so that armed conflict may be avoided.

THE FLUIDITY OF INTERNATIONAL RELATIONS

If we had to use one word to describe the current situation in Northeast Asia, it would have to be fluid. Over the past several years, important changes

119

have taken place. The Soviet Union has consistently built up its military forces in the Far East over the postwar period, but the pace of that buildup was particularly striking in the past 15 or so years. Under General Secretary Gorbachev's leadership, however, the Soviet Union has withdrawn its troops from Mongolia and has taken other significant steps in improving its relations with China. China has also determined to improve its relations with the USSR, and the process of normalization between these two regional powers culminated in the first Sino-Soviet Summit in more than 30 years in May 1989. Even on the Korean peninsula, which has been the focus of East-West confrontation in the region, a North-South dialogue has commenced, and South Korea has opened economic relations with China, the Soviet Union, and East European countries.

Again, it would be overly naive to assume that this fluidity promises a rosy future of peace and stability. Many factors of instability remain. The Korean peninsula remains characterized by military confrontation. It is there that the interests of the three nuclear powers—the United States, the USSR, and China—meet. In addition, local conflicts, such as in Cambodia, continue in Southeast Asia, and in the Philippines and Burma the potential for political and economic instability is high. The threats posed to Asian stability are diverse; they differ from country to country, and by locality.

Rather than conceiving of the various military deployments in the region as a balance, it is more appropriate in Asia to describe them as a mosaic. Even a small conflict carries the potential for widespread changes. Thus it is necessary to be careful in undertaking any changes in the existing security system. Also, the level of economic and social development in Asia differs by country as well as by region, and, therefore, a European-style bipolar structure, based on conflicting values, that separates the "free world" from the "Communist world" cannot be replicated in Northeast Asia. Alliances and friendly, as well as adversarial, relationships can be very fluid in this region. Strong nationalism, particularly in those countries that have experienced colonization, further enhances this fluidity. Therefore, it is important that Western cooperation also be flexible. The stability of Asia is not based simply on a military balance, and thus political, economic, and social stability must also be considered.

International relations are interwoven; a change in one element invites movement in others. The entire system may be destabilized by even a minor shift in policy. Particularly in times of transition, foreign relations must be structured within a clear framework. Without such a framework, confusion and disruption may result. Therefore, it is important for the United States and Japan to jointly support the current security system, centered around their alliance, while incorporating flexibility to the changes currently underway in international relations. In Asia, a political dialogue should be initiated among the various conflicting powers, and the level of military confrontation should be reduced as much as possible to allow for the introduction of measures to reduce tensions.

EVALUATING "DEFENSIVE" MILITARY POLICIES: THE JAPANESE EXAMPLE?

One of the most significant potential changes in international relations in recent times is the Soviet Union's adoption of a "defensive defense" policy. Announcements of this shift in the Soviet military force posture attracted a great deal of attention in the West, and the likelihood and consequences for Western security should be carefully evaluated. It is still too early to conclude that the Soviet Union has indeed truly begun to change its military doctrine, but an assessment of how to recognize this change is called for.

In Asia, the signs of change are not as conspicuous as in Europe. The Soviet Union has over the past decade or so conducted a steady buildup of its forces in the Far East, and it is difficult to interpret their developments as "defensive" in nature.

Prior to examining Soviet military behavior in the region, however, it is necessary to have some standard of assessment for what a defensive military strategy would look like and how a force posture could be designed to reflect defensive intentions. Because there is no agreed-upon standard for discriminating between offensive and defensive forces, I present Japanese policy here as an example of an attempt to design such a force posture, and then examine Soviet behavior in light of the example.

Article 9 of the Japanese constitution asserts that Japan will not maintain "war potential" and denies Japan the "right of belligerency of the state."[1] In the postwar period, therefore, Japan has maintained an "exclusive defense" posture. The Japanese Defense Agency explains this policy as follows:

> [Japan's] "exclusively defense-oriented" [policy] refers to the posture of a passive defense strategy that is consistent with the spirit of the Constitution. Thus, adhering to this policy, Japan will initiate its defensive operations only when its land is attacked by a foreign power or powers. Even then the scope of military operations and the level of the defense forces to be mobilized will be kept to the minimum required for the purpose of self-defense.[2]

Clearly, the definition of "the minimum required for the purpose of self-defense" changes in accordance with the international environment and with the standard of military technology. However, the Japanese government has determined that the possession of offensive weaponry capable of imposing massive destruction on an adversary's territory surpasses the scale of "the minimum required" force level. In other words, Japan interprets this to mean that it will not maintain ICBMs, long-range strategic bombers, or offensive aircraft carriers.[3]

Moreover, the right of self-defense will be invoked by Japan only under the following three conditions (officially referred to as the "three requisite conditions"): (1) when there has been a sudden and illegitimate act of aggression against Japan, (2) when there is no appropriate means to deal with this aggression other than resort to the right of self-defense, and (3) when the use of armed force is confined to the necessary minimum.[4] The government also maintains the position that the Self-Defense Forces will not be dispatched "to foreign territorial land, sea, and airspace for the purpose of using armed force," since that would surpass the minimum level for self-defense.[5]

In addition, with reference to nuclear weapons, the Japanese government abides by the "three nonnuclear principles" of not possessing, producing, or permitting the introduction of nuclear weapons into Japanese territory. In 1976, Japan ratified the Nuclear Nonproliferation Treaty and committed itself not to produce or acquire nuclear weapons.

The structure of Japan's defense capability, therefore, reflects this concept of exclusive defense. This capability has been viewed as part of a policy set comprised of the U.S.-Japan alliance system and the defense efforts of Japan's Self-Defense Forces (SDF). The foundation of the current defense program, the National Defense Program Outline that was adopted by cabinet resolution on October 29, 1976, emphasizes that the objective of Japan's defense policy is "deterring invasion" through the maintenance of a Japanese defense capability and through U.S.-Japan security cooperation. In response to attack, Japan's Self-Defense Forces will "repel limited, small-scale aggression" and if the scale or mode of aggression exceed the capability of the SDF to do so, then the SDF will act in cooperation with the United States.[6] Thus the objective of Japan's defense forces is the effective management of a limited and small-scale aggression.

In November 1978, the United States and Japan agreed to the "Guidelines for U.S.-Japan Defense Cooperation," and according to these guidelines, the Japanese Self-Defense Forces are primarily responsible for defensive operations for the Japanese homeland and surrounding sea and air territory. The guidelines clearly assign responsibility for offensive operations to the U.S. military.[7]

Japan's defense conception can be summarized as follows:

1. For nuclear deterrence, Japan will rely on the United States.
2. The Self-Defense Forces will defend Japan against a small-scale and limited attack.
3. A larger attack will be met by the defensive operations of the Self-Defense Forces and the offensive operations of U.S. forces.

The complementarity of U.S. and Japanese military capabilities—Japan's Self-Defense Forces and U.S. military forces deployed in the Far East—based on this conception of defense is clearly visible. The Ground Self-Defense Forces are for the defense of the Japanese homeland and do not maintain the supply and logistical capability for overseas deployment. Counterattacks against an invasion will be implemented through cooperation with the United States. The Maritime Self-Defense Forces also have a defensive mission: the defense of Japanese ports and the three straits, and an antisubmarine warfare capability for the seas around Japan. For naval power projection, Japan depends on the United States. The Air Self-Defense Force is comprised of air defense interceptors, radar systems for air warning and control, and antiair missile groups. The ASDF does not hive the long-range capability to attack the adversary's bases.

If we look only at Japan's defense capability, then it is an "exclusive defense" posture. In fact, the structure of Japanese forces could be considered as a realization of the Soviet's conception of a "defensive defense" posture.

SOME QUESTIONS REGARDING THE SOVIET UNION'S DEFENSIVE DEFENSE POLICY

In his Krasnoyarsk speech on September 16, 1988, General Secretary Gorbachev proposed seven items regarding security in the Asia-Pacific region. He promised that the Soviet Union would not introduce any new nuclear weapons into the area and called for the United States and other nuclear powers to adopt a similar policy. Similarly, he proposed a freeze on naval deployments in the region and called for consultations on reducing military tensions. Then, on December 7, 1988, speaking at the General Assembly of the United Nations, he made public the Soviet Union's intention to reduce the number of Soviet troops by 500,000 over the next two years. At the Sino-Soviet Summit on May 17, 1989, Gorbachev stated that 120,000 of these troops would be cut from forces deployed in the Soviet Far East. This reduction represents approximately one-fifth of the total Soviet troops in the Far Eastern Military District and the Transbaykal Military District (597,000).

Progress is being made in the negotiations to reduce U.S. and Soviet strategic arsenals, and we can expect that these negotiations will bear fruitful results. With regard to tactical nuclear forces, it will be necessary, as it was with the negotiations on intermediate-range nuclear forces (INF), to take a global approach. An approach that treats tactical nuclear forces on a regional basis, separating the European theater from the Far Eastern theater, has little meaning given the potential to move nuclear weapons quickly from theater to theater. Also, negotiations on controling chemical weapons will have to be conducted on a global basis.

Although Soviet arms control proposals for Asia are welcome, there are several issues to be raised regarding the Soviet assertions of "new thinking" about its foreign relations and of the adoption of a defensive defense policy.

First, the Soviet proposal to freeze naval deployments in the Asia-Pacific ignores the geographical asymmetry between the United States and the Soviet Union. The Soviet Union as a continental power can link its European region to its Far Eastern by overland transport. The United States is linked to Japan and Korea by sea lanes, and the maintenance of this linkage is vitally important for the Western countries. Imposing limits on naval forces in the Pacific would have little influence on the Soviet Union's military capability, but for the Western allies it would have a critical impact.

Second, the content of the defensive defense policy articulated by the Soviet Union is vague. Even if the USSR emphasizes homeland defense based on this conception of a defensive defense policy, it produces a serious problem for Japan. The Soviet Union defines its goal as the creation of a zone of reconnaissance and observation and of "sea denial" to Western naval forces out to 2,000 kilometers from its coast. Furthermore, in waters closer to the Soviet Union, it would like to create a region of "sea control" that would keep Western naval forces from approaching Soviet waters. The Japanese islands are very close to the Soviet Union and are in fact included in this "sea control" region. Thus, even if the Soviet Union argues that it has adopted a defensive defense policy, that very same policy has an offensive character when seen from Japan.

Also, there is no doubt that the Soviet Union is emphasizing the Sea of Okhotsk as a region for the deployment for its SLBMs. In the case of a conflict, it is highly likely that the Soviet Union will conduct combined sea and land operations against Japan in order to secure the Soya Straits, the main route between the Sea of Japan and the Sea of Okhotsk. Since 1987, the Soviet Union has been deploying ground troops on Japan's Northern Territories (the four islands including Kunishiro and Etorofu under dispute between Japan and the USSR), very close to Hokkaido. At present, these troops are being strengthened and now are on the scale of one division. This buildup is a threat to Japan's security.

Third, if we apply, for example, Japan's conception of exclusive defense to the Soviet conventional forces deployed in the Far East, the following issues emerge. (For the purpose of this argument, I use the data on Soviet Far Eastern deployments provided by the *Pravda* interview with USSR Defense Minister Yazov published on May 28, 1989.)

The Soviet Union would need to abolish from its air forces the long- and midrange bombers that are capable of attacking neighboring countries. The Tu-16 (*Bagder*), Tu-22 (*Blinder*), Tu-26 (*Backfire*), and Tu-95 (*Bear*) aircraft are all forces with the objective of a strategic attack, and thus they present a threat to surrounding countries. In the Soviet data released in May, the number of tactical aircraft was given as 1,690, and within this total, a large number of attack aircraft, such as the Su-24, (*Fencer*) and the MiG-27 (*Flogger D*), were included. Thus the Soviet Union would need to make large reductions in its air forces to make its claim of a defensive defense policy valid in the Far East. It should be noted that the U.S. Air Force deploys no more than about 50 F-16s at Misawa Air Base in northern Japan.

Also with regard to naval forces, the Soviet Union has 55 major surface ships deployed in the Pacific, including two aircraft carriers and 46 attack submarines. Furthermore, the Soviets have greatly increased their coastal defense capability. In addition, the Soviet navy's aircraft, the Tu-22M (*Backfire*), presents a major threat to the SLOCs between the United States and Japan. Thus, if a strictly defensive posture was to be devised in the Far East, the Soviet Union would need to make broad reductions in its surface fleet, its submarine fleet, and its Tu-22Ms. For comparison, it should be noted that the U.S. Pacific Fleet deploys a total of 35 attack submarines, and the Japanese Maritime Self-Defense Forces have 14 diesel submarines. Moreover, the Soviet Union's deployment of naval infantry, designed for combined sea and land operations, in the vicinity of Japan is indeed revealing and cannot be construed as consistent with a defensive defense strategy.

The number of Soviet ground troops deployed in the Far East is about 600,000. Even if a reduction of 120,000 troops, as proposed by Gorbachev, is implemented, this still leaves 480,000 troops in the region. If we also consider that the USSR deploys 12,600 tanks, 14,300 military vehicles, and 16,400 pieces of major artillery, Soviet forces in the Far Eat have a strong offensive character. American ground forces deployed in the Far East include one division in South Korea and the Marines (less than one division) deployed in Okinawa. Japan's Ground Self-Defense Forces maintain no more than 150,000 troops, 1,200 tanks, 620 armored vehicles, and 650 howitzers. Again, the Soviet Union would have to make major reductions in the number of its troops and in its equipment deployed in the Far East if it were to adopt a defensive policy along the lines of Japan's concept of exclusive defense.

Fourth, the Soviet Union has frequently put forth the proposal of devising Confidence-Building Measures (CBMs) in the Asian region. Before we jointly consider CBMs, however, there are measures that the Soviet Union must take unilaterally. The activity of Soviet aircraft and the Pacific Fleet have become very pronounced in the area around Japan. Soviet aircraft make more than 320 activities around Japanese airspace annually. In December 1988 a Soviet Tu-16 bomber violated Japanese airspace over Okinawa where the United States has an air force base. In recent years, the Soviet Union has directly flown over Japanese Air Self-Defense Force radar stations, and the number of flights considered as training exercises for attacks on these radar sites has risen. The activities of the Pacific Fleet reveal a similar pattern; 625 Soviet ships transit the three Japanese straits, the Tsushima, Tsugaru, and Soya straits, each year. In August 1988, 25 ships, including the missile cruiser *Frunze,* entered the Sea of Okhotsk through the Soya Straits to conduct exercises. Prior to any negotiations on regional CBMs, therefore, the Soviet Union must halt this type of activity, which generates regional concern and belies Soviet intentions for tension-reduction.

Fifth, it will be necessary for the Soviet Union to increase the transparency of information on its military. Although the total amount of Soviet military expenditure has been announced, the Soviets should go further and clarify the content of that total in a manner similar to the Western countries. The USSR should also make public information on its force posture, including the number of troops and major weapons, and their deployments. Moscow should give serious considerations to these types of measures.

SINO-SOVIET RAPPROCHEMENT AND DEFENSIVE DEFENSE

Gorbachev visited China in May 1989 and met with Chinese leaders, including Deng Xiaoping and Li Peng, ending 30 years of conflict between the USSR and China. State and party relations were both normalized, and the two countries agreed to cooperate in economic and cultural affairs. In addition, they agreed to establish a consultative group on reducing military forces deployed along their common border and agreed to an early withdrawal of Vietnamese troops from Cambodia. Gorbachev proposed that the USSR and China make the border a demilitarized "region of peace."

China places a priority on economic construction, and for that purpose, emphasizes the maintenance of a peaceful strategic environment. Clearly, for China, Sino-Soviet detente facilitates economic construction by reducing the threat from the north. China argued that there were three obstacles to rapprochement with the Soviet Union: (1) Soviet withdrawal from Afghanistan, (2) the withdrawal of Soviet troops from the Sino-Soviet border and a resolution of border issues, and (3) the withdrawal of Vietnamese troops from Cambodia. Gorbachev completed the Soviet withdrawal from Afghanistan in early 1989, promised the withdrawal of Soviet troops from Mongolia, and also promised to resolve outstanding border issues with China. Even with the remaining "obstacle" regarding Cambodia, the two countries reached an agreement on an early withdrawal of Vietnamese troops during the summit meeting.

Therefore, we can say that the three obstacles to an improvement of Sino-Soviet relations have been virtually resolved. However, China and the Soviet Union share a 7,000-kilometer-long national border, and they have a history of confrontation. It is difficult to believe that the conflict between them has been completely abolished. Even if Sino-Soviet rapprochement has been achieved, it is unthinkable that the alliance relationship of the 1950s will re-emerge.

Undoubtedly, the Soviet Union and China would both benefit from the adoption of a more defensive military configuration along their mutual border. The onus of change in this respect has rightfully been on the Soviet Union, which has had a superior military capability. In fact, China's strategy of a "people's war" essentially has the characteristics of a defensive strategy. The People's Liberation Army, equipped with outmoded weaponry, cannot match the modernized Soviet forces. Thus the "people's war strategy" is a strategy of engulfing advancing Soviet troops in a "sea of people." The Chinese capability of crossing the border and invading the Soviet Union remains limited. Furthermore, China has only a minimal capability of returning a nuclear strike on Soviet territory if the Soviets launched a nuclear strike against the PRC, and this limited Chinese nuclear capability has served only as a deterrent to Soviet nuclear attack.

In order to promote economic construction, the People's Liberation Army reduced its troops by one million, and it is reported that a further cut of 300,000–500,000 troops is planned. China also receives American military assistance, but relative to the Soviet Union's equipment modernization program, China remains far behind. For this reason, China undoubtedly welcomes a reduction of Soviet Far Eastern forces and the adoption by the USSR of a defensive defense policy. Moreover, until China succeeds in its program of economic construction, it has little choice but to continue to rely on its "people's war strategy."

Over the long term, the question of how China considers the development of its naval forces will be an important issue for Asian security. In September 1988, China succeeded in launching a guided missile from a submarine. The Chinese may be considering the development of an SLBM force to counter Soviet nuclear forces, and an issue for the future will be whether or not the Chinese navy will be satisfied with a coastal defense fleet.

THE POTENTIAL FOR CONFRONTATION ON THE KOREAN PENINSULA

There is little doubt that the most dangerous region in Northeast Asia is the Korean peninsula, where over one million North and South Korean troops face each other across the demilitarized zone. Each side distrusts the other, and both sides continue to modernize and increase the strength of their forces. If a conflict erupts, it is clear that the scale of death and destruction will rapidly expand.

The United States and the Soviet Union have a common interest in not inviting a military confrontation in Northeast Asia. If the United States and the USSR think about peace in the region, they should start with Korea, the vortex of East-West confrontation. Both the United States and Soviet Union are committed by treaty to support their respective allies. The PRC also has an alliance relationship with North Korea. The eruption of conflict on the peninsula would seriously affect relations between these three nuclear powers.

For this reason, it is important to maintain the military balance between North and South Korea. Since the 1970s, North Korea, which devotes 20–25 percent of its GNP to its military budget, has markedly increased the strength of its forces. The North Korean army currently has 32 divisions, including 3,200 tanks and approximately 750,000 troops. Its navy is mainly comprised of 20 submarines and 8 high-speed missile ships, with a total tonnage of 60,000 tons. The air force maintains 740 tactical fighter aircraft. In terms of ground forces, the army is about 1.4 times the size of that of South Korea, and it is said that in terms of mechanized forces and firepower, such as battle tanks, armored carriers, and automatic guns, North Korea is superior. In recent years, the North has made efforts to modernize its forces, and it has received fighter aircraft from the Soviet Union, including MiG-23s, Su-25s, which have a superior antiground attack capability, and MiG-29s, which have a superior air control capability. North Korea has also received SA-5 surface-to-air missiles.

South Korea considers this increase in North Korean military capability as a serious threat and is making efforts to improve its own military forces, investing 5.5–6 percent of its GNP each year in defense. It has an army of 540,000 with 21 divisions composed of the three forces deployed between the DMZ and Seoul. Korea is also buying American TOW antitank missiles and antitank helicopters. The air force has approximately 380 aircraft, mainly F-4s and F-5s. In 1986 Korea began to introduce F-16s and is strengthening its early warning system so that it can better prepare to meet a surprise attack. In addition, the United States currently has about 44,000 troops deployed in Korea, including one infantry division and several air wings, and a U.S.-ROK Joint Command was established in Seoul. The Korean military strengthened its military capability in the 1980s, and as long as the current rate of economic growth continues, it is likely that in the 1990s the current imbalance between North and South will be corrected.

The deterrent function of U.S. forces in Korea is crucial to peace and stability on the peninsula, however. The stationing of forces in Korea is an important symbol of America's commitment to defend South Korea. The influence of this symbol of American commitment is immense. If U.S. forces in Korea were completely withdrawn or greatly reduced, it would give the North Koreans reason to believe that the United States had rescinded its commitment. This misunderstanding could be very dangerous.

That is not to say that there is no potential for reducing tensions on the Korean peninsula. Recent efforts in that direction have been promising, and a multilateral dialogue, in addition to North-South talks, would be an effective means of reducing the potential for conflict there. On October 18, 1988, after the Seoul Olympics, President Roh Tae Woo made his first speech at the General Assembly of the United Nations. In it he made two suggestions. First, he proposed that a North-South Summit should be held in order to promote the reconciliation and unification of the two Koreas. There should be no preconditions that the summit discussions would include a declaration of nonaggression or the no-first-use of force. Second, Roh proposed that a consultative organization be established for a discussion of how to resolve Korean problems that would include the participation of six countries: North and South Korea, the United States, the USSR, the PRC, and Japan.

North Korea immediately rejected these proposals and blamed South Korea for attempting to gain international recognition by perpetuating the division of the two Koreas. China supported the North Korean position by calling for a three-power conference that would include North and South Korea and the United States. The Soviet Union, while not referring directly to the North Korean statement, suggested its support by stating that it would not neglect the North Korean position on the issue. From the perspective of regional peace and stability, however, a political dialogue on the Korean peninsula should include not only the two Koreas, but also other concerned countries. There is a good chance that President Roh Tae Woo's proposal for a political dialogue, which takes into account the flexibility of both the Soviet and Chinese positions of late, will be realized.

North and South Korea are struggling with the process of reinaugurating bilateral talks. This is one means of easing tensions on the peninsula, and clearly the beginning of a dialogue at the initiative of the two Koreas is desirable. However, the mistrust that has developed from a long separation and confrontation is very deeply rooted. Therefore, the United States and the Soviet Union should continue efforts to consult with their allies while trying to implement practical steps toward reducing tensions on the peninsula.

Specifically, the first step should be to examine what we would call preliminary measures. North and South Korean participation should be requested in arms control measures adopted by the United States and the USSR. Both East and West powers should recognize the two Koreas, and both countries should be admitted to the United Nations at the same time.

Then, as a second stage, confidence-building measures should be considered. There is already a "hot line" between Seoul and Pyongyang. Also, since 1986, when the United States and the ROK conduct their annual joint "Team Spirit" military exercises, they have informed North Korea and have indicated that they are willing to accept North Korean observers. This attempt to implement measures such as the mutual exchange of information on military exercises, the mutual exchange of observers, open access to military information, and the mutual withdrawal of troops from the DMZ should be continued. And, then, as a third stage, the mutual reduction of troops can be considered. These types of tension-reduction measures should be pursued slowly, in a step-by-step manner.

Efforts to reduce tensions on the Korean peninsula should be conducted at the initiative of the two Koreas. However, it is possible for the four powers deeply involved, the United States, the USSR, China, and Japan, to participate in the talks and to cooperate in measures designed to reduce tensions as proposed by President Roh.

Of course, the Korean peninsula should not be the sole focus of efforts to reduce tensions in the region. But if steps can be implemented to bring stability to the peninsula, the outlook for arms control in Northeast Asia will be positive. The scope of application of confidence-building measures can expand from the highly volatile Korean peninsula, and the road to arms control in Asia could potentially be broadened to include other countries in the region.

THREE PRINCIPLES FOR REDUCING TENSIONS IN NORTHEAST ASIA

Because of the uncertainties surrounding the Soviet Union's defense policy, the United States and Japan must continue to make the defense efforts needed to meet Soviet deployments in the region. However, this does not mean that we should ignore the potential for reducing tensions.

Tensions in the region are multiple—they exist both at the level of the major powers and at the regional level. Thus efforts to encourage a reduction of tensions must take into consideration the peculiar characteristics of Asia. Three principles should be followed.

First, to the extent possible, those issues that stem from relations between the major powers and those that stem from regional relations should be separated. Individual solutions to these individual sources of tension should be devised. For example, the Soviet call for reducing SLCMs is an issue that should be resolved from the perspective of the U.S.-Soviet global nuclear balance, and the regional powers should keep out of these negotiations. In recognizing this principle, U.S. allies such as Japan and the ROK can avoid becoming entangled in the complexities of nuclear issues. Likewise, the resolution of problems on the Korean peninsula and of the Cambodian conflict should not be linked to other issues.

Second, negotiations should be pursued gradually. The Soviet Union and North Korea have repeatedly proposed across-the-board arms reductions to be taken in one step. However, many questions remain regarding Soviet and

North Korean military deployments. With the facts unclear, it is impossible to undertake discussions on arms control. The primary reason that Gorbachev's proposals and the North Korean proposals cannot be taken seriously is that there is no indication of how these goals are to be achieved— there is no practical prescription for implementation attached to them.

Implementation of arms reductions in Asia, particularly in Northeast Asia, will need to be approached in a step-by-step manner. For example, although releasing information on deployments, giving prior notice of military exercises, and announcing the transfer of troops are minor aspects of arms control, a certain level of trust can be built through these efforts and the basis for future mutual arms reductions can be improved.

Finally, as a third principle, an emphasis should be placed on *economic and technological cooperation in Asia*. It is clear that economic and technological exchange with the West is desired by both the USSR and North Korea. To a certain extent, this will accompany progress in reducing tensions, and economic cooperation should be considered in tandem with political talks. Japan has a great deal of influence in this respect, and close coordination between the United States and Japan will be necessary.

Notes to Chapter 6

1. *Defense of Japan 1988* (English ed.), p. 74.

2. *Ibid.* This White Paper, originally published by the Defense Agency, Japan, has been translated into English by The Japan Times, Ltd. with the permission of the Defense Agency for wider distribution.

3. *Defense of Japan 1988* (English ed.), p. 75.

4. *Ibid.*, p. 75.

5. *Ibid.*

6. *Ibid.*, p. 87.

7. *Ibid.*, pp. 86–87.

Chapter 7

Nonoffensive Defense and the Korean Peninsula

Håkan Wiberg

Korea is one of the areas in the world that has undergone the most radical transformations in terms of strategic position and significance, as well as in many other ways.

In conventional economic terms, Korea was for a long time highly underdeveloped. The half-century of Japanese occupation, notwithstanding anti-Korean oppression, also brought some amount of industrialization. After the utterly devastating war in 1950–53, the country lay in ruins, and in both parts of the peninsula, reconstruction had to start from very close to zero. Since then, there have been two economic miracles, although based on quite different development strategies and with equally different results. As a result, both states in the peninsula are among the industrialized nations today, both have had very impressive long-term growth since 1953–and not only in GNP and similar terms. If we look at indicators of how *people* actually live, such as life expectancy and infant mortality, both of them have rapidly moved into the group of societies with a high welfare level.

Although it lies outside the topic of this chapter to inquire into how this has been brought about, some speculative hints should be made. Some factors are probably common, like the "Phoenix factor" and the "Confucian factor." The former[1] signifies that the very devastation may contain seeds to future growth, both in terms of motivation and in terms of getting a fairly modern production equipment in economically strategic sectors, substituting that which was destroyed. The latter phrase refers to the observation that—under certain circumstances—the cultural factor of Confucian ethics and social philosophy also seems able to serve as a powerful motor in economic development.[2] It does not seem to matter much whether this basic Confucian pattern is combined with, or overlayered by, other bodies of thought, such as Christianity, Buddhism, Liberalism, or Marxism-Leninism.[3]

A third factor appears to be what, if we look for an abstract formulation of a common denominator, we may describe as a high degree of state intervention in the economy.

There the similarity ends, however: the two Koreas have "out-Europed Europe" in quite different directions. In the North, we find a more autarchic (*juche*) and a purer command economy than anywhere in Eastern Europe (except, perhaps, Albania). In the South, the economy is probably *domestically* more (Manchester) liberal than anywhere in Western Europe, the state intervention largely being outward-oriented and having to do with a considerable degree of protectionism, as well as giving (authoritative, even if often unofficial—and hence usually needed) advice concerning export coordination, patterns of technology import, and so on. This advice is given by state bureaucrats who tend to be well trained, patriotic, and not very corrupt.[4] South Korea belongs (together with the other Confucian nations and a few more) to the fairly limited number of success stories for export-led development based on vast foreign loans for investment.

THE DIVISION

Korea was a united nation for about a milennium,[5] mostly under a (usually fairly nominal) Chinese suzerainty.[6] After the defeat of China, the Japanese also quickly rebuffed a Russian attempt to get a foothold, and Korea then suffered a half-century (until 1945) under a Japanese colonialism that worked hard—and failed completely—at virtual ethnocide, trying to "Japanize" Korea entirely and producing, in fact, a nation that was more intensely Korean than ever.

During World War II, Korea seems to have been very peripheral in the notions of the Allies concerning the postwar political order. There was agreement that Korean sovereignty was to be restored, but seemingly no more concrete plans. The administrative division made between the United States and the USSR in 1945 appears to have been improvised at short notice, using the Japanese administrative map as a point of departure,[7] and it may well initially have been intended as precisely what it claimed to be: provisional. For another couple of years, there was (at least nominal) agreement in the United Nations on the creation of an undivided and sovereign Korea,[8] even if the great powers had frustrated Korean nationalism by insisting that this sovereignty be postponed until after a five-year period of great power administration. (As it happened, only the United States and the USSR remained among those that were originally intended.)

Whatever the original intentions of the great powers, the ensuing processes were largely echoes of the polarization taking place in Europe, and the net result in 1948, after separate elections in the two parts—the legitimacy of which were challenged by the other side—confirmed the original fears. Instead of one Korea, we got two Korean states, the *Republic of Korea* and the *Democratic People's Republic of Korea,* each one proclaiming to be the only legitimate one. They had a very great number of skirmishes with each other after the military withdrawal of the superpowers from the peninsula. After the outbreak of the hot war two years later, each of the two states suffered virtual extinction as such in 1950–51. By late summer 1950, there was only the Pusan pocket left under ROK control, before it was saved by U.S. military intervention (nominally under the UN flag). A few months later, only a slice of land along the Yalu River remained under DPRK control, before it was saved by Chinese military intervention (nominally consisting of volunteers).[9] Finally, after three years of war, the material devastation of most of the peninsula and a number of dead well exceeding a million, with even more wounded and homeless, the armistice negotiations in Panmunjom resulted in truce lines that differed rather little from the status quo ante. It also resulted in two Korean states that are far more intensely suspicious of each other (and each other's friends), as well as investing considerably larger proportions of their wealth in military preparations than the two German states ever were, or did.

The division betweem them is also very much deeper than the German division ever was, whether you look at it in terms of trade, family unifications, travels, mail communication, mass media, or whatever. Reunification talks have now gone on intermittently for almost two decades, but the results—also in all these partial terms—have so far been virtually nil, at least if you judge them by German standards.

Whether one regards this as a cause or as a symptom, the political philosophies behind the unification approaches by the two Korean states have been maximally at odds with each other. The southern approach has essentially been a "functionalist" one,[10] proposing a long list of smaller common measures in terms of trade, communications, practical cooperation, and so on, and preferring to postpone any wider measures until one has seen the results of the smaller ones. The northern approach has essentially been "legalist," proposing a loose confederation of the two states, which are to keep their own socioeconomic systems.[11]

It does not take much social imagination to see that there has been very little in the proposals of one state that has had much attraction for the government of the other state.[12] At the same time, it would be facetious to write the negotiations off as purely ritualistic (however much ritual they have contained). They have had a very powerful background: the strong all-Korean feeling of belonging together, however detestable the *government* of the other state is shown to be in one's own political line and mass media, a feeling that is too strong not to pay tribute to in terms of holding out unification as a goal and a possibility.

THE MILITARY POSTURES

We can see a built-in paradox, if we compare the desire for unification with the defense planning of the Korean states. On the one hand, both of them have a level of military preparedness that is very high by international comparison and that is legitimized by the suspicion that the other side might try to achieve unification on its own terms—by force—and should not be led into such a temptation.[13] On the other hand, precisely these high levels of military preparedness can be assumed to contribute to there being very heavy odds against any Korean unification, even one based on friendly agreement between the two states.

The reason for this assertion is as follows. Let us consider what would be the result of a hypothetical peaceful creation tomorrow of a Korean unitary state, a creation that must then have taken place with the consent of both governments. It could therefore hardly be anything else than nonaligned. In this case, the situation, as looked at through the lenses of security planners of the surrounding states, would change drastically. Before the unification, one had a Korean peninsula, which was dangerous both because of the bitter enmity between the two governments and to the extent that the amassing of so many men in uniform and so much military hardware creates a risk for explosion in itself. At the same time, however, those two armed forces are of approximately the same (great power) size and may therefore be expected to balance each other for some time, even if it would come to a hot war, except in the very unlikely case (given their mutual wariness) that one of them would manage to surprise the other one.

A unified Korea, however, would be something very different, unless there were a very considerable degree of disarmament before or during the reunification process. If we add the armed forces of the two states, we would get a picture (in quantitative terms) of the fourth largest standing army in the world (perhaps even the third, depending on how one counts reserves), having a tank fleet and an air force surpassed by at most half a dozen states each. These crude *quantitative* measures of specific dimensions will mean an overestimation, not taking quality, compatibility, and such into account. Even if we go to an (in itself equally crude) overall measure as defense expenditures, however, a united Korea would rank somewhere around the tenth place in the world. In case of a Japanese-Korean conflict (the possibility of which would, under these hypothetical circumstances, hardly be dismissed by Japanese defense planners), it might be able to pose a serious military threat to Japan. In case of a new Sino-Soviet conflict (which the defense planners on neither side would exclude as impossible), the party that might manage to drag Korea in on its side would thereby have made the threat to the other side much more serious.

One can therefore easily imagine four states that can be expected to have very little interest in the unified Korea seen in this scenario. Given the strong anti-Japanese sentiments that are still found in Korea, Japan would be expected to object to having such a vast military force, with no superpower restraint, as its closest neighbor. Soviet and Chinese security planners would—in line with ordinary worst-case analysis—be expected to put greater emphasis on the possibility that Korea joined the opponent than on that of its joining themselves. And the United States would have no interest in Korea leaving its alliance for an unpredictable nonalignment. If these four states would have an interest in Korea *not* getting unified, it appears very likely that they would be able to block such a development, whatever lip service they pay to it. For the two present Korean states to be able to ignore the political, economic, and such sanctions that might emerge, they would have to see a very strong common interest in some yet to be invented formula for unification that they could agree on. And for this reason, it is most likely that the precondition for any qualified Korean unification to be feasible is that it is preceded by a rather drastic reduction of offensive capability—whether by disarmament or transarmament—on both sides, down toward internationally more normal levels, and that negotiations to that purpose between the Korea states would have to be part of any realistic negotiations about unification.

Such negotiations will be difficult, given the mutual fear and given the strategic geography and the structure of the forces. The problem with the geography is the lack of strategic depth, as clearly demonstrated by the war in 1950–53, where both states were very close to getting completely overrun and were only saved by external intervention. That, in its turn, means that both sides have force structures that are highly offensive *in the technical sense,*[14] and for both of them, the main option for defense is seen in having own forces that permit strong counteroffensives—and that, for that reason, can also be used for offensive purposes.

Let me put this in a wider perspective. If a hypothetical unified Korea endowed with the added military capability of the present Korean states would appear problematic to *neighbors,* each of these two states is also bound to appear problematic *to the other one.* It is hardly an exaggeration to say that almost *any* state having as a neighbor a state with the military capability of either of the Korean states would feel insecure unless it (1) *either* were clearly stronger itself (which, as indicated, very few are), (2) *or* had very good relationships with it (which the Korean states very obviously do not have with each other), (3) *or* felt that it had a credible deterrence arrangement, whether by own rearmament or alliance formation.[15]

Both Korean states have opted for the first part of the last option: a high level of armament. Traditionally, they have also looked for an alliance with a major power to shore the deterrence up, whereas today it is only South Korea that has a firm military alliance, with the United States, whereas North Korea does not,[16] and has therefore put an even greater share of its resources than South Korea into military preparedness.

The result of this is the mentioned spiral of offensive capability. In itself this is quite understandable, if we look at it through traditional strategic lenses. The lack of strategic depth of either state has to do with both of them being quite small in terms of area.[17] It is therefore difficult for either of them to avoid doing precisely the things that increase fears on the other side, however unintended this may be. The important question then becomes whether "difficult" also means "impossible."

NONOFFENSIVE DEFENSE?

There are thus several symptoms in Korea that seem to indicate Nonoffensive Defense (NOD) as a solution:

1. There are clear tendencies to an arms race in the peninsula. In the last decade, both states have rearmed at a pace that clearly exceeds the global average. As far as can be judged from historical statistics,[18] this means that if there is a military confrontation, the risk that this escalates into a hot war is higher than for most other pairs of neighbor states. In addition, it means that both Korean states are deprived of resources that could have been used for development.[19]
2. These problems with crisis stability and escalation stability are further exacerbated by the fact that the force compositions of both states are fairly offensive, as indicated, e.g., by the sizable tank and air fleets.
3. This combination of two armed forces with strong offensive features also makes possible disarmament negotiations difficult, and this, as we have argued, will in its turn constitute one more obstacle for Korean unification.

It is easily seen that all three problems would be ameliorated by the Korean states transarming in the direction of NOD. It would be expected to have the net effect of counteracting the arms race (since the defense of one side would be seen as less of a threat to the other side), improve crisis stability (by there being less to gain by striking first), and improve the possibilities for disarmament negotiations (there being less to lose by the possible results).[20]

It is, however, much more difficult to define a concrete *therapy* than to make the *diagnosis*. For given the problems that the strategic experts in either state are bound to see with the other state, the leeway for purely or mainly military/strategic solutions is in any case rather limited. Neither of the Korean states has the option of getting so much stronger than the other one that there would be no need to fear it. North Korea fears would obviously be reduced by South Korea leaving the alliance with the United States, or at least reducing its military presence and the extent of the joint military exercises,[21] but there is little likelihood of either the United States or South Korea contemplating this, until and unless there has been a sizable reduction in North Korean military capability. And whereas reducing this capability is one obvious thing that North Korea could do to reduce fears in South Korea, this is not likely either, as long as North Korea sees itself facing a very strong South Korea—with the resources of a superpower to draw on in addition.

The remaining option would then be a NOD option, trying to break the spiral of offensive capability by converting the military forces in more predominantly defensive directions, so as to reach a situation where either state is defensively superior to the other state. In slightly more concrete terms, this might be done by creating military structures with credible capability of *blocking* possible attacks from the other side, but considerably less capability of *attacking* it. This would mean, among other things, to counter tanks and air forces with antitank and antiair missiles rather than mainly doing so by own tanks and air forces; to put more resources into a territorial defense by attrition in the border areas on either side of the DMZ, and less into counterattack capability.

To try to work such a scenario out in more detail would require both more military expertise and, in particular, a much more detailed acquaintance with the strategic geography and military systems of the Korean states than the present author possesses. The main points that can be made in general terms are the following ones:

1. For the reasons mentioned, already the military factors indicate that it will be more difficult in Korea than in Germany to work out plans for Nonoffensive Defense that have the possibility of being sufficiently acceptable to security planners to have any chance of realization. For that reason, it should be seen as a challenge.
2. The linkages that we have described between political and military matters indicate that any process of moving toward Nonoffensive Defense in Korea has to be combined with a process of solving political problems between the two states—and at least some of these problems have to be solved before it can have much possibility.

Hence, we also have to look at the political dimension.

THE POLITICAL DIMENSION

There are a number of political problems defining obstacles to Korean unification and even to Korean detente. As we have seen, some of them derive directly from the military capabilities and postures of the Korean states. Others, however, do not to the same extent have military roots.

One of them concerns the very particular position of the United Nations in the Korean peninsula. Whereas in many other conflicts the organization has been able to play helpful roles as mediator or as truce observer, in Korea it appears prevented from so doing by its position as a *party* to the conflict. This is a heritage from 1950, when some very particular circumstances[2] came to cast it into that role.

In addition to this, the constellation of historical experiences and the present character of the conflict makes it quite difficult to find any other significant actor that could appear as mediator. If we assume having diplomatic relations with both sides to be a minimum condition, already that one rules all the major powers out.

This, in its turn, seems to indicate that a cross-recognition might be a significant step forward toward some kind of political solution.[23] In other words, when the United States and Japan have established diplomatic relations with North Korea at the same time as China and the USSR do so with South Korea, there will be more roads open; and since some of the nonrecognitions are clearly carried out in solidarity with other nonrecognizers, rather than expressions of own national interests,[24] this problem may be near its solution.

At the same time, one should not overestimate how much closer it would move by that multiplication of roads. For it is not self-evident that the Korean states *need* any mediator, nor that a mediator could do much to solve some of the specifically Korean problems.

One of these problems consists in both of them claiming *Alleinvertretungsrecht,* to borrow a term from earlier German postwar history: the exclusive right to represent the Korean nation. As long as this is the case, at least one—and possibly both—of them will continue to oppose cross-recognition, the legitimizing argument for that position being that this would be tantamount to recognizing the division of Korea.[25]

Another problem lies in the hierarchical character of Confucian ethics. The crucial problem may not so much be the existence of two socioeconomic systems that are very different, at least on the doctrinal level; for that might be solved by creating a loose confederation preserving both under the umbrella of common—but mainly formal—institutions. These institutions in themselves, however, will create problems. If we have some federal assembly, even if a highly formal one, the problem will arise whether it should be composed on the basis of a 50-50 principle or in correspondence with the numbers of inhabitants of the present states. This problem, however, may be possible to resolve by a formula that grants each state veto power against unacceptable decisions, whether directly or in terms of a rule about qualified majority. Even so, the problem of the top of the pyramid will remain: who is to be the president, or chairman, or whatever title is created, of the hypothetical federation. Here, it appears unthinkable that one of the states would accept the other one to supply the incumbent of that office, so more complex formulas will have to be invented: cochairmanship,[26] or some system with rotating chairmanship, which can also be found in some constitutions; but the particular Korean solution will have to be a matter of Korean inventiveness.

Notes to Chapter 7

1. Also observed in some of the most devastated areas in Eastern and Western Europe after 1945.

2. Other examples are seen in Taiwan, Hongkong, Singapore, some periods in modern China, the Chinese minorities in several countries, the Shinto-Confucian Japan, and Japanese minorities abroad.

3. In the Japanese case, we should rather speak about Shintoism overlayered by Confucianism.

4. The last point should be specified in the light of the incumbent government usually accusing the preceding one—and the opposition the incumbent government—of widespread corruption. The point is that the corruption that occurs appears to be largely "inside the (Korean) family" and therefore giving little access to unwanted foreign influence.

5. See William E. Henthorn, *A History of Korea* (London: Macmillan, 1971).

6. Another matter is that there is long tradition of factionalism in Korea. See Woo-Keun Han, *The History of Korea* (Seoul: Eul-Yoo Publ. Co., 1969).

7. If the USSR had had any desire in August 1945 of occupying all of Korea, rather than making an agreement with the United States and waiting for their troops to arrive in the South, there would have been little to stop it.

8. As late as 1947, there was still unanimity in the UN on the principle of restoring Korean unity and independence, but later in that year, the work of the U.S.-Soviet commission definitely broke down. Ironically, one of the issues was that the USSR wanted to ban those Korean political actors that had opposed the U.S.-USSR condominium and insisted on immediate Korean sovereignty.

9. Apart from arms deliveries and some use of air force close to the Soviet border, the USSR did not intervene militarily.

10. See, e.g., Chung-Won Choue, *Towards Korean Integration* (Seoul: Kyung Hee University, 1986).

11. See, e.g., *Statement of the Committee for the Peaceful Unification of the*

Fatherland, Pyongyang, October 14, 1988. For a theory-based analysis of Korean integration, see Johan Galtung, "Divided Nations as a Process: One State, Two States, and In-between," *Journal of Peace Research*, 9, no. 4 (1972): 345–60.

12. The measures proposed by South Korea will be seen in North Korea as designed to break its self-defined autarchy, thereby giving South Korea one-sided influence in the North. The constitutional measures proposed by North Korea will be seen in South Korea as designed to create a political platform for one-sided influence in the South.

13. Given the history during the lifetime of the present leaderships, it would be surprising if they did not harbor such suspicions.

14. Here, it is not a matter of whatever *intentions* they may have—it is well known that they suspect each other of the worst. When we describe the force structures as offensive, the point is that they are clearly *capable* of delivering deep strikes into the territory of the other side.

15. This formulation, it should be underlined, is about what leaderships of states *tend to think;* it does not take a stand on whether such thinking is in some sense *true*.

16. Another matter is that it may take into account the traditional Chinese position, demonstrated in 1950, that China will not tolerate all of Korea dominated by any power that is seen as hostile, but then, it may seem unpredictable in Pyongyang exactly what would be seen in Beijing as fulfilling this condition.

17. NATO has traditionally been worried about the lack of strategic depth of West Germany and Western Europe. The area of North Korea is half, and that of South Korea two-fifths, of that of West Germany.

18. See Håkan Wiberg, 'Arms Races—Why Worry?' in Nils Petter Gleditsch and Olav Njølstad (eds): *Arms Races: Technological and Political Dynamics* (London: Sage, 1989), pp. 32–57.

19. According to *SIPRI Yearbook* and *The Military Balance*, South Korea spends more than 5 percent and North Korea some 10 percent of its GNP on military preparations. (Other sources, although probably less reliable, put the proportion in North Korea as much higher, up to 25 percent). It must be conceded that both Korean states have performed economically better than most other states with such military burdens, but these nevertheless represent a very sizable untapped development potential.

20. Systematic theory cannot be presented in this short chapter. For some essential points, see chapter 1 by Albrecht von Müller in this volume. The development of theory has essentially taken place in the FRG. For the most thorough presentation of the West German debate in English, see the book by Bjørn Møller; *Resolving the Security Dilemma in Europe: The Debate on Non-Offensive Defense in the Federal Republic of Germany*, (1990).

21. Judging by North Korean behavior in negotiations, these vast annual exercises lie very high on the list of what the North sees as threatening. The South Korean fears, in contrast, primarily tend to add the USSR to the threat picture. See, e.g., Jong Yul Yoo, *Soviet Military Build-Up in the Far East: A Threat to Regional Peace* (Seoul: Kyung Hee University, 1984).

22. Especially the heavy Western dominance in the UN at that time combined with a period of USSR absence—and hence no veto—in the security council.

23. In fact, it has been suggested by both Korean states but at different time points—and when one was for, the other one was against.

24. For example, I would assume China and South Korea, the USSR and South Korea, and Japan and North Korea to be mutually interested in diplomatic relationships.

25. As the German example shows, that argument tends to be forgotten, once the two states have nevertheless found it practical to have direct contacts.

26. For example, modeled on the earlier U.S.-USSR cochairmanship in the Eighteen Nation Disarmament Committee.

Part IV

Defensive Strategies in Regions of Conflict: The Middle East and Central America

Chapter 8

Defensive Defense in the Middle East

Alex Gliksman

The concept of defensive defense first gained prominence in West Europe during the early 1980s as NATO approached the deployment of Pershing IIs and Ground-launched Cruise Missiles. Offered by peace researchers in West Germany as a conventional alternative to that nuclear deployment, defensive defense was initially more a vague notion than a well-defined military concept. To the degree that the skeleton was given flesh, it was perceived by traditional defense analysts as largely a defense avoidance scheme. In place of traditional armies, there would be territorial forces, including militias in which every man and woman would take on invading tanks and troops, using locally stored stocks of rifles, mines, and antitank weapons. Even today, some continue to view defensive defense as a citizen's self-defense plan.

But as thinking and analysis have progressed, many of the concept's proponents have now come to recognize that the military environment is far more complex and demanding than many of the early ideas implied. Trained elite troops are likely to overwhelm numerically superior citizen-soldiers, however valiant their effort. Absent centralized command, control, and intelligence, defensive resources are unlikely to be at the right place, at the right time, to scuttle an invasion by tanks, self-propelled artillery, and other offensive capabilities. Mobility is an equally important element, not just to seek out the invader but also to deprive him of an easy target for preemptive attack.[1]

None of this is likely to come cheap. This will certainly be a disappointment to those who have looked to defensive defense as a means of sparing national budgets for social services rather than for the military. Indeed, the kinds of defensive (nonprovocative) conventional capabilities required to discourage conventional aggression could very well require more rather than less defense spending. As a first observation, conventional defenses of whatever type are likely to be costly. It is helpful to remember that when NATO first turned to the deployment of battlefield nuclear weapons in the 1950s, finances, not the desire for nuclear options, were the dominant consideration. Time and again since then many of the best-laid NATO conventional arms plans have fallen to the budgeteer's ax.

Moreover, there is a close connection between the defensive defense as it has evolved and the process of treaty verification. Former President Ronald Reagan's favorite words to Soviet President Mikhail Gorbachev, "trust but verify," seems equally applicable to both types of regimes. Whereas arms control seeks to deny benefits that might otherwise accrue to a potential violator, defensive defense seeks to produce a negative return for a country or group of countries contemplating aggression. In each case the objective is to discourage the violation or aggression from happening in the first place by restructuring the playing field in a way that there is no advantage in upsetting the status quo.

Similarly, the capabilities required for sustaining defensive defense and arms control will likely incorporate closely related and in some instances the same technologies. The basic parameters of an effective defensive defense and of an assured treaty compliance system are comparable. The former requires timely warning of a surprise attack; the latter requires early warning of preparations for arms control breakout. The former requires arms capabilities favoring the defender over the aggressor; the latter requires having programs (such as research and development in relevant fields) that would provide an effective hedge should a treaty violation occur. If the ideal can be achieved, the system should work in a way that the victim would come out ahead. If that could be arranged, then there would be no incentives for misadventure.

These parallels between requirements of arms control and defensive defense provide further reason to believe that defensive defense will not be obtained on the cheap. A hint of this is to be found in the cost of implementing the Intermediate-range Nuclear Forces (INF) Treaty. American officials responsible for the verification required in implementing the treaty claim that capabilities required and consequently the costs involved in monitoring the accord dramatically exceeded their expectations, some even suggesting that living with the treaty may cost more than living with the missile. Here, it is useful to note that the budget of the U.S. On-site Inspection Agency represents only part of the cost of the treaty regime. These same officials predict a major leap in costs for verifying any accord that is likely to emerge from the Vienna talks on conventional armed forces in Europe. Keeping track of conventional capabilities is an exceedingly complex problem. There are far more forces to watch than was the case for the INF accord. The fact that large numbers of armaments will be permitted under an agreement means that monitoring cannot be reduced to the either/or problem of the INF "double zero" approach. Given that defensive defense is also conventional in nature suggests that a similar rate of growth in implementation costs can be expected.

Without belaboring the parallels between the prerequisites of defensive defense and arms control much further, it is important to observe that both compliance and nonaggression are highly dependent on effective intelligence, as the above implies. Intelligence is what verification and early warning are all about. Effective intelligence is built on technologies that can provide warning to the violated side and a framework that rapidly and unambiguously exposes misbehavior so as to assure that the defender can take timely action in marshalling military resources.

As the above indicates, in addition to providing the capabilities and a framework that will deny to the aggressor the advantage of surprise, defensive defense also requires a force structure that denies an aggressor his objective(s) if he, nevertheless, attacks.

Denial and punishment are closely related. As much as we might wish otherwise, effective military capabilities are designed to inflict punishment. Wounding an aggressor is a key aspect in preventing him from successfully executing his military plans. It is doubtful that defensively oriented military strategies and structure can do away with this. A defensively oriented system should also be designed with the sorts of incentives that would lead a potential opponent to opt for a military structure that will reduce his offensive potential. It also requires the sort of constraints that minimize the risk that a defender will mistakenly use his forces to effect.

Creating a concrete foundation that short circuits efforts to achieve surprise in an attack and that denies an opponent the means to conduct large-scale offensive operations is the key to an effective defensive scheme. Architects of such arrangements have sought to use a combination of geography and technology in putting a defensive defense framework into effect, independent of the specific region. Political circumstances are even more important. For absent the will to collaborate, even the best conceived defensive defense scheme is unlikely to move beyond the drawing board.

The use of zones is a common feature of most defensive defense arrangements. These seek to reduce prospects of successfully initiating a surprise attack. They do this by prohibiting offensively capable forces from operating in forward areas—that is, at potential points of contact between the forces of the opposing sides. At the same time, both sides are encouraged to deploy rugged defenses in these forward zones; such forces typically include anti-tank weapons, mine-laying equipment, and other obstacles of various types. Although tanks, mobile artillery, bridging equipment, and other forces defined as offensive may be permitted to operate in zones to the rear, defensive defense seeks to encourage reductions in such war-making potential. Mobile reserves that can counter offensive maneuvers are encouraged by some, including Soviet military writers.

But offensive weapon-free zones in and of themselves are unlikely to significantly reduce surprise attack options. Capabilities to spot preparations for and movements by prohibited forces into areas of disengagement are equally important. These early warning assets can take a variety of forms, ranging from airborne radar surveillance to strategically located ground-based acoustic, infrared, and optical sensors and to manned observation posts. It is vitally important that a combination of assets be used and that a consultative mechanism be created to manage disputes. Such a mechanism could enhance transparency and predictability if it is also used to routinely exchange military data and discuss military plans and programs. This helps avoid reacting prematurely in the face of false alarms or unintended infractions. Should the alarm be real, an interlocking set of early warning sensors can help the defender accurately characterize the attack and better deploy his forces in response.

The combination of well-conceived zones and effecting monitoring systems can reduce the potential of achieving surprise. Clear lines of demarcation, indicating where military units may and may not operate can also act to enhance the capabilities of even the most advanced technology sensor systems. Data lends itself to a variety of interpretations. Increased data-gathering capabilities will not ease this situation. In some instances, knowing more about the other side's military activities might compound problems and increase risks of misinterpretation. If being overwhelmed by data results in a failure to take timely defensive action or in a precipitous defensive response, the consequences could be disastrous. This is especially true during crisis. The downing of an Iranian airliner by the U.S. Aegis cruiser *Vincennes* during the Iran-Iraq war is an example.

Well-structured arms control zonal arrangements creating clear demarcation lines could help enhance early warning and mitigate risks that military transparency will needlessly heighten tensions that could inadvertently lead to war. This is becoming increasingly important as many of the newly emerging systems designed to monitor military activities in peacetime and provide early warning in the event of war are double edged. Once an attack has begun, these peacetime eyes and ears could provide the data required to track and target the offending forces.

Denying a potential aggressor the prospect of successfully launching a surprise attack is only half the problem; the other half is denying an aggressor the wherewithal to conduct large-scale offensive operations. There may be instances where an aggressor might strike absent surprise if he believes he can achieve his objectives by the use of brute force. In the 1973

Arab-Israeli war, Egypt's hope of victory rested more on attrition than on achieving surprise.

Having the staying power to conduct large-scale military operations relies on a forward-based combat support infrastructure. Defanging a military force's offense-sustaining capabilities can be achieved by several methods, including restrictions on and the destruction of combat support capabilities. Air bases and logistic support installations located close to the likely lines of engagement should be subject to such constraints. Without fuel, munitions, repair facilities, and such, even the best-planned offensive would quickly run out of steam.

How does this all apply to the Middle East? What follows is little more than a sketch that seeks to measure the requirements of defensive defense against conditions in the region. Its bottom line suggests that whereas it would be extremely difficult to overcome the profound hurdles to the wholesale application of defensive defense in the Middle East, appropriately modified and selectively applied variants might have a limited role to play in the region.

Applied to Israel and its neighbors, Israel's geography and small population are a problem. Zones of disengagement are generally impractical, given Israel's lack of geographic depth. Zones have been successfully applied in the Sinai, where geographic depth and a will for political compromise, not evident in Tel Aviv's relations with any of its other neighbors, lent themselves to creating an effective mutual nonprovocative security regime, where offensive forces and their infrastructure have been successfully disengaged.

As a study conducted for the Canadian government points out, third party involvement by the United States and by the United Nations was critical to the process.[2] According to this study, American involvement helped to spur Egypt and Israel to cooperation at each stage in a difficult, protracted, and incremental process. Washington played an invaluable role as designers and providers of verification systems and technology, including giving Egypt an independent surveillance capability required to overcoming fears of being caught unprepared for an attack. This provision of technology helped meet a low tech Egypt's need for equality in verification means while giving assurance "that neither side would be better off—or at least not worse off—by adhering" to the agreed zones.

Reliance on not one but on a multidimensional set of interlocking verification assets involving third parties, including the United Nations, and the potential belligerents was an another important ingredient. This helped to provide a more transparent environment under which disengagement could occur. Among other things, both sides flew reconnaissance aircraft in controlled zones while the United States operated a ground-based early warning system in the strategic Mitla and Gidi passes and conducted various forms of overhead reconnaissance, which it shared with both sides.

In the Sinai case, both side's desire to positively alter their security relationship was a critical motivating factor. In the other successful, currently operating Middle East zonal arrangement, the Golan Heights demilitarized zone separating Israel and Syria, the elements of mutual fear and hostility and domestic politics joined to create this arrangement. In contrast to the Sinai, the Golan Heights has little relationship to defensive defense. The outcome of the 1973 war in the Middle East, this UN-supervised demilitarized zone provides a face-saving device for the Syrians, who despite their words fear an Israeli attack on Damascus, as well as for the Israelis, who worry that the Golan Heights could be used for rocket and terrorists attacks or, more seriously, as a launch pad for a Syrian invasion of Israel's heartland. Nevertheless this "fig leaf" arrangement, as one knowledgeable official termed it, provided a tripwire that helps somewhat defuse each party's concern for the other's potential offensive capability, while avoiding an appearance of weakness in the eyes of their respective domestic political constituencies.

There is a need to identify various possible zones in the Middle East. There are plenty of conflicts besides that between Israel and its neighbors. In many of these other cases, such as Iran-Iraq, Syria-Iraq, Lebanon-Syria, Egypt-Libya, and indeed, Libya with most of its neighbors, geography is not an issue. There is enough depth in these nations to create effort zones. But more than anything else, it is the politics of conflicting interests and ambitions that is an obstacle to making the concessions required to even discussing, to say nothing of creating, nonoffensive security arrangements.

In brief, the region in whole or in part seems to be largely without the political ingredients required for applying a defensive defense system or systems. As a first step, one needs to identify the precursors that will build confidence and will make for the kinds of effective arms constraints that can make defensive defense possible in the future.

Confidence-and security-building measures (CSBMs) developed for use in Europe and in managing U.S.-Soviet security relations would seem to be relevant. Measures such as the prior notification of exercises and maneuvers would help create the predictability in military conduct that builds confidence. This process would also benefit from the avoidance of threatening activities such as massing military forces at or near border areas. Setting ceiling on the size of military exercises could also form part of a package of CSBMs designed to avoid incidents. Making military activities transparent should be part of confidence-building efforts. The exchange of military plans and program data and visits, including exchange of observers at military exercises, could have a positive role in advancing this goal. Jointly manned crises management centers would help deal with threats, both real and perceived. This may be especially useful in the Middle East where non-state actors might seek to promote their interests by creating incidents that provoke tensions between states in the region. The involvement of the United Nations and of states from outside the region that can serve as honest brokers with technical expertise required to monitor events could be an invaluable part in such arrangements.

Among other things, this will require gaining the concurrence of nonstate players who can contribute to or undermine international arrangements. This involves national, religious, and ethnic groups and subgroups that have their own stakes in the region, including the Palestinians, Shiite Moslems, and many other groups and factions. As is demonstrated on a daily basis by the Lebanon conflict, failure to address their interests could undermine both government-to-government (and intragovernment) understandings that might otherwise prove practical.

Notes to Chapter 8

1. For a review of the evolution of the thinking of defensive defense proponents, see David Gates, "Area Defense Concepts: The West German Debate," *Survival* (July/August 1987), pp. 301–17.

2. Brian S. Mandell, *The Sinai Experience: Lessons in Multimethod Arms Control Verification and Risk Management* (Ottawa: Department of External Affairs, 1987).

Chapter 9

Nonoffensive Defense Strategies in Central America

Luis Herrera-Lasso

This paper was written in September 1989. Since then, numerous events have taken place all around the world. In Eastern Europe changes have led rapidly to an unexpected point of no return. In Central America, in December of 1989 the American troops based in Panama with additional forces and weapons coming from the United States, invaded the country, took Noriega prisoner and gave way to the government of Guillermo Endara who was the main contender of Noriega's candidate and presumably the winner of the presidential elections of May. In February 1990, Violeta Chamorro, the candidate of the *Union Nacional Opositora* (UNO) defeated Comandante Daniel Ortega in the presidential elections of Nicaragua, an astonishing outcome, at least for most foreign observers.

The two main changes in Central America, that can be interpreted as major victories of the United States—in the case of Nicaragua as their first victory of a low intensity war and in Panama as a result of the most primitive use of force in international relations—have had already great consequences for the subregion. For the first time in more than a decade, presumably there will be, in the seven countries of the subregion, governments with a more homogeneous political perspective and, the most important to the United States, rather compatible with U.S. perspectives and interests in the area. At the same time, this trend tends to isolate the FMLN in El Salvador and is a significant setback for Cuba. There are still problems and conflicts: the internal struggle in El Salvador; the demobilization of the contras and the enormous problems that the new government of Nicaragua surely will face; the internal and external consequences of the invasion for the current Panamanian government; the signs of an increasing conflict in Guatemala and, perhaps the most delicate, the near future of Cuba.

The United States has recovered some of its lost diplomatic and political support in Latin America—to a good extent due to the economic crisis—and has managed rather successfully to lead hemispheric attention mainly to the problem of drug trafficking and production. Despite this, it seems that political and military victories are barely consolidated in the economic realm. Experts argue that in Nicaragua as much as in Panama, the population supported Guillermo Endara and Violeta Chamorro because it was impossible to survive the low intensity war lead by the United States. Therefore, these populations have a significant expectation of an economic reward coming from the north. This view does not seem to be shared in Washington, or if so, to a very lower extent. In Colombia, Bolivia, and Peru, the economic aid in exchange for adopting the American strategy in the drugs terrain, has meant very little help to solve the critical internal situations.

Most analysts today share the conclusion that the world is passing through a major process of accommodation that requires major regional and sub-regional adjustments in which all actors involved have to play an active role. That is happening in Eastern Asia and in Europe. In this hemisphere the process, if any, seems to be extremely slow. What has been considered the victory of the cold war by the United States, has tended to reinforce the perspective that they can act as a superpower reviving the concept of the "zone of influence" far from considering other actors in the hemisphere as partners. This delay in perceptions could have irreversible consequences not only for the Latin American countries but also for the United States.

The appearance of nuclear weapons gave way to a new era in which offensive strategies became practically the only choice. Nevertheless, a strategic balance based on offensive nuclear capacity has posed many questions in the minds of politicians and strategists. If the balance fails, mutual assured destruction appears as the inevitable end of the conflict; in other words, surrender and annihilation are the only possible options.

For the main actors involved, this reasoning led them to explore "defensive" defense systems. In the nuclear realm, this took the form, among others, of the Strategic Defense Initiative (SDI), which pretends to base nuclear deterrence much more on a defensive strategy than on an offensive one.

To Western European countries, the vulnerable position of nuclear hostage has given way to historic debate centered on the need for alternative defense strategies that make it possible to get rid of nuclear weapons through effective defense systems that increase the cost of the attack for any potential aggressor. The agreements to dismantle intermediate nuclear weapons—SS-20, Pershing, and Cruise missiles—reflect this common interest. This strategic approach also has clear implications for conventional war strategies.

Additionally, we also find interesting proposals and historical cases of civilian-based deterrence strategies that have proved to be effective and not simply idealistic. Such has been the case of India and its nonviolent resistance against the British Empire, or Norway during the World War II, when a civilian nonmilitary resistance against the Nazis was developed that proved rather successful.

Thus the exploring and development of alternative defense strategies are not only a result of the lack of a reasonable option in a scenario of war with nuclear weapons, but also a result of conviction, already proved in some cases, that it is possible to have an effective defense policy not necessarily based on the use of force.

In Latin America there is neither the problem of nuclear weapons based in the region nor that of being a nuclear hostage, nor is there a foreseeable scenario for a major conventional war. There are, however, some other realities absent in Europe, such as territorial disputes, internal conflicts, deep economic crisis, and an inter-American system with more questions than answers, that lead us to defense strategies different from those known in other parts of the world.

To center the discussion, I start with some general features of the inter-American system as a framework for the defensive strategies of Latin American countries. In the second part of the chapter, I focus on the Central American scene where we have faced actual military conflicts for the past decade or more.

This chapter does not pretend to be an exhaustive analysis of the causes of conflict in the subregion. Many works from different perspectives have been written on the subject. The focus here is on the different strategies of defense and survival of these small actors on the international stage, those with very limited power and resources to consolidate a national project and to develop a real capacity to preserve and defend it in a world where they must deal with global and hemispheric interests out of their own control.

Finally, we discuss some ideas about the security and defense agenda of Latin America as well as its interrelation with security and defense perceptions in other regions of the world.

THE INTER-AMERICAN SYSTEM: OLD STRUCTURE AND NEW REALITIES

The political-strategic reality of Latin America has to be analyzed in a wider scope: the hemispheric or inter-American system as characterized by the presence of a dominant actor, the United States. This geopolitical condition has two main consequences: first, there is a structural limitation for all countries of the system, except for the dominant actor, to assume hemispheric roles. This situation leads us to a first conclusion: not one of these countries has any possibility, real or potential, to build up hemispheric offensive strategies.

Second, the system's center of power has a magnetic force over all the actors of this geographical area that influences significantly their foreign policies and their strategic and defense concepts. In this context, bilateral relations with the United States become the dominant variable of all the international interactions of the Latin American actors.

This phenomenon has internal and external implications. Facing the outside world, the possibilities of the use of force as an instrument of foreign policy for Latin American countries is limited to certain restricted scenarios: to be used in the subregional spaces for expansionist purposes, or to be used as an instrument of their defense policies to deter military aggressions coming from potential enemies.

In the first scenario, we can only talk of one country with the capacity for a sophisticated strategy with clear geopolitical aims, which is Brazil. With this exception, most countries have a very low potential to engage in an offensive war. Territorial disputes, many of them historical, appear as the only cause that could eventually lead to an offensive war. Nevertheless, an historical survey of the past 40 years gives us an extraordinarly low level of actual military conflicts among Latin American countries, especially compared with other regions of the Third World.

The internal consequences of the geopolitical conditional as a result of relations with the dominant actor have not been less significant. On one side, the presence of a dominant hemispheric actor has also contributed to avoiding conflicts between regional actors and extraregional powers. On the other side, the existence of military or civilian authoritarian regimes, with national security concepts that very frequently have led to repression, counterinsurgency, and a surpassing of the empire of the law by the empire of authority, are to a greater or lesser extent interrelated with the ideological and strategic approach of Communist containment coming from the United States as a global policy.

New Hemispheric Realities

Since 1979 there have been some significant changes within the inter-American system. In May of that year, the initiative of the United States at the Organization of the American States (OAS) to deal with the internal conflict in Nicaragua only a few weeks before the victory of the revolution was rejected by most Latin American countries. The war in the South Atlantic islands of Malvinas/Falklands, in which the United States took the side of

Great Britain, was a second serious strike to the inter-American system. In this context, Latin American countries involved in international disputes started sending their claims to the United Nations while the OAS entered into a period of inactivity, very close to economic and political bankruptcy.

We also witnessed in this decade the emergence of the Contadora Group as a joint effort of four Latin American countries to contribute to a peaceful solution in Central America. This was possible due to four main circumstances: (1) the "vacuum of power" left by the inadequacy and ineffectiveness of the OAS, (2) related to the first, the inability of the United States to lead the OAS as had happened during the three previous decades and the apparent loss of U.S. moral and political leadership in the region, (3) the increasing international activism of the major countries of the region after two decades of growth and stability, especially in the three major countries of the Caribbean basin: Mexico, Venezuela, and Colombia, and (4) the interest of the actors involved in isolating domestic and subregional conflicts from the East-West conflict approach.

In this evolving process of the Inter-American system, we must also consider the increasing importance given to subregional approaches to local problems and conflicts, a global trend reflected in the Contadora and Esquipulas schemes, as well as in the efforts toward a major cooperation among Brazil, Argentina, and Uruguay in the south of the hemisphere. The Group of Three by Colombia, Mexico, and Venezuela seems also to fit in this trend.

CONFLICTS AND WAR

Central America cannot be isolated from hemispheric and global trends. Notwithstanding, we have to consider at least the following particularities of the subregion:

- The internal political scene in each country.
- The gravitation of the dominant external variable in the internal and subregional process.
- The increasing relevance of the subregional approach to deal with internal and subregional problems.

The internal political situation is rather different from one country to another. Despite the undeniable similarities among theses nations in terms of common history, language, and culture, today each one seems to be in a very different stage of political life. In Nicaragua there is a new political system with a new regime coming from a revolution. El Salvador is still in a dark period of civil war that has lasted for almost a decade. Guatemala, and specifically the governments of this country, has succeeded in defeating the armed opposition in the military field but has not been as successful in solving the problems that gave birth to those movements despite democratic forces that have been gaining ground in the past several years. Panama seems to be approaching a critical chapter in its internal and external political life.

Honduras, the poorest country of the area, survives due to the external aid received from the United States, playing in exchange an active supportive role to the American policy toward the subregion, but without signals of real improvement in its political, economic, and social life. Costa Rica, the country with the largest tradition of stability and democracy in the area, has played an ambivalent role trying to preserve political neutrality and economic development in a complex subregional environment. Belize is trying to overcome economic crisis while attempting to consolidate its political and social structure as a new state.

Facing this complex scenario, the United States has adopted a very active role. As a result of a global doctrine of Communist containment, its strategy has focused on overthrowing the revolutionary regime in Nicaragua, preventing the victory of the armed opposition in El Salvador, and maintaining the political sympathy and support of countries without internal conflicts such as Costa Rica and Honduras. The aim of actively joining Panama seems to face more difficulties than expected, creating a new scenario of conflict that can bring unpredictable consequences for the entire region.

The unilateral and unipartisan policies of the United States over the past eight years or so have resulted in the so-called strategy of low intensity war, with simultaneous actions on indirect intervention in the economic, military, political, and diplomatic fields, which has prevailed over diplomatic or negotiated solutions toward the achievement of American aims. According to the "rollback" doctrine, results are still far from what was expected. The Sandinistas are still in power, the FMLN has not been defeated, there have been approximately 150,000 dead in the subregion, concentrated in El Salvador and Nicaragua, and about three million people have migrated from outside their homes, their countries, or the subregion, many going to Mexico or to the United States. Additionally, the subregion as a whole is in the worst economic and social situation of its recent history.

All efforts made by other actors of the region or outside it—governmental and nongovernmental—to bring peace to the area have clearly been insufficient, although they have served to avoid a worse situation and to alleviate, to a certain extent, the deteriorated economy of these countries. Nevertheless, one decade after the turmoils were initiated in that part of the world, real solutions still seem far out of reach.

DEFENSE STRATEGIES OF THE CENTRAL AMERICAN COUNTRIES

In general terms, we can talk of four different real or potential threats to the security and national interest of Central American countries:

1. Internal foes.
2. Aggression from neighboring countries.
3. Intervention of the United States.
4. The Communist threat.

At the first level, armed opposition in some of these countries has become a threat to the internal security, the stability of the system, and the political group in power. In El Salvador, the opposition, using a strategy of guerrilla war, has rather successfully resisted years of counterinsurgency strategy led by the government with significant military support from the United States. In this context, the civilian population has very limited means to protect their lives, homes, and families, and to make an honest living within the minimal levels of survival. A climate of insecurity prevails in many areas of the country.

According to the government, the main enemy is the FMLN, a subversive guerrilla movement with Communist support. This perception is shared and supported by the United States. Aggression from neighboring countries is not an issue. To the FMLN, the main enemy is the authoritarian government, supported by the armed forces, which has historically maintained a closed political system that favors the oligarchy's economic interests against those of most of the population.

In this complex case, the strategies of the most active military and political actors are diverted to the internal conflict. Their strategies are offensive and the objective is to defeat the adversary in the military field. The military stalemate that has lasted for several years has resulted in a pre-eminence of military strategies over negotiation schemes with very small success in the realm of political bargaining between the different forces involved.

In the subregional scenario, the participation of the Salvadorean governments has usually served to back the positions of the United States, although its formal participation in the Esquipulas efforts shows historical concern for a common destiny that can be crucial for the future of these countries. If we look at the wide web of economic interactions among these states, we can support the structural validity of this assertion.

In Nicaragua since 1982 we have a defensive strategy focused on the protection of national territory against external aggressions. The civilian population plays a very significant role in the defensive strategy. On one hand, there are 200,000 people in the militias, with specific tasks in the defense strategy against the counterrevolutionary forces and, if the moment comes, they are trained to face a direct external intervention. On the other hand, the defense strategy has diverted much of the limited economic resources to the war effort. The Sandinistas' attention is centered on the evolution of the political process—to have an open political system but to preserve the revolutionary regime—and on the recovery of the economy. The defensive strategy is maintained, but expectations of a war have decreased significantly. First, because of the victory over the armed opposition, the "Contras." Second, because they have increased the cost of a direct military intervention to a level that makes this option unlikely, at least for the near future.

Nicaragua is also very much involved in the subregional military balance. Its defensive strategy is designed to face external aggression coming from the United States or from its neighboring countries with U.S. support. Its military buildup is designed mainly for these purposes, based on defensive weapons such as tanks, artillery, and antimissiles systems, with a poor aviation strategy, inadequate to fulfil major offensive missions. Nevertheless, the government of Nicaragua has a much superior capacity of military mobilization than any of its neighbors. It is also true that a powerful defensive strategy can be converted, to a certain extent, into an offensive war effort.

Therefore, it is understandable that the problem of the subregional military balance arises in all area negotiations. To Nicaragua, its military strategy is a defensive one against external aggressions, but the government claims not to have any intentions to attack its neighbors. Therefore, a military balance vis a vis each country is not necessary. In contrast, Honduras, El Salvador, and Costa Rica—to a lesser extent Guatemala—consider that Nicaragua is a military threat to the rest of the countries in the region. Thus the balance should be posed in the following terms: Nicaragua on one side and the rest of the countries on the other.

Theoretically, this approach would let all countries but Nicaragua increase their military capabilities. For the Sandinistas, the core of the problem is the asymmetry of forces vis a vis the United States and, the only solution, an evidence of nonhostile intentions toward Nicaragua from the superpower. In this scenario, military balance among subregional countries should be a much easier matter to deal with. Supporters of this approach contemplate on the other side the necessary withdrawal of any other military support to the region, including those coming from Cuba and the Soviet Union.

Honduras has become a special case. Traditionally the main perceived threat comes from El Salvador. Despite this, Hondurans have developed, with the support of the United States, a defense strategy that can also be effective in facing potential aggressions from Guatemala or Nicaragua. Additionally, and apart from the national military buildup to be used in potential conflicts with neighboring countries, Honduran territory has been used to host the armed groups fighting the Sandinistas and temporary military bases of the United States. Without an internal conflict, in real terms Honduras is the country with the highest indirect involvement in military operations in the area. The agreement of the government to remove the Contras from its territory is a sign that even for this country, which relies very much on American economic aid, the internal and external costs of this involvement have become too high.

Costa Rica is in a rather different situation. Despite its deteriorated economy, it has preserved a stable political system. To face the turmoils, Costa Rica has pretended to maintain a neutral position, although it has played an active role in the diplomatic sphere, with a conservative approach, and its territory has been used by the Contras to fight the Sandinistas. Costa Rica does not have a proper army, but it has increased considerably the number of police and other paramilitary forces. As in Honduras, Costa Rica relies very much on the military protection of the United States in the case of a military aggression coming from Nicaragua. It is not likely that Costa Rica would initiate aggression against Nicaragua.

During the first half of the 1980s and despite the economic deterioration and frequent political changes, Panama managed not to get involved in the subregional turmoil, but it is a member of the Contadora Group. Since 1985 the situation has changed. Apparently the reluctance of the Panamanian governments, and especially of the Defense Forces of Panama, to cooperate in the subregional strategy of the United States has led to another very sensitive scenario. The differences between Panama and the United States toward the subregional crisis and concerning the accomplishment of the Torrijos-Carter agreement have complicated the internal and external situation of this country. Since June 1987 the United States has engaged in an unsuccessful strategy to remove the chief of the armed Panamanian forces and these forces have been working internally to prepare the population for a military resistance against the potential aggressor. Panama could become the center of the crisis in the subregion at any moment.

Guatemala is also a rather different case. With the strongest army in Central America, the governments supported by the armed forces have defeated the guerrilla movements. The Guatemalan counterinsurgency strategy has been one of the most successful on the continent. Additionally, the military, besides having a strong tradition of nationalism and independence, has also managed quite well without significant military support coming from outside.

Externally, Guatemala does not foresee any significant threat. Aggression from Honduras or El Salvador is very unlikely. With Mexico, relationships have improved significantly with the democratic government of Vinicio Cerezo achieving good progress in sensitive issues such as the refugees. Although there are historical prejudices about Mexico, Guatemala does not expect aggression, and it seems very unlikely that it would take the initiative in this direction.

The position played by the Guatemalan governments in the subregional crisis has been rather interesting. First, they have acted with great independence from the United States. Second, they were quite cooperative during the years of the Contadora process, due in part to their interest in having a good relationship with Mexico. Third, they have been very active in the subregional initiatives—Esquipulas—with an interesting influence as mediator in the crisis. It is also interesting that during all these years of crisis, they have not had any trouble with their neighbors.

In the case of Belize, we also find a very different scenario. Because of the historical territorial claims of Guatemala, Belize maintains a British military force in its territory. Although the force is there only to protect Belize's independence, this is a foreign military presence in the subregion. There is uncertainty about the future of these forces, including the possiblity that they will be replaced by forces of the United States.

In most cases, defense strategies are not, and cannot be, offensive ones in this region. Nicaragua has a defensive strategy with an extensive participation of the civilian population. Panama seems to be working in the same direction. In these two cases, the expected aggression would come from outside the subregion because none of the other countries by themselves would have any chance of winning an offensive war. El Salvador and to a lesser extent Guatemala are much more concerned with internal conflict and counterinsurgency strategies, in the first case, with a crucial participation of an external actor. Honduras and Costa Rica would have very little chance in a military confrontation against Nicaragua unless they join an offensive war with a direct intervention coming from abroad. Belize has not shown the slightest interest in getting involved in the subregional crisis.

The Communist threat, one of the main elements of the conflicts in the area from the perspective of the United States, seems to be less important since significant advances between the two superpowers—very much due to *perestroika*—could lead to an explicit acknowledgement that Communist threat in Central America does not compel the United States to become a major participant in subregional conflicts. There is wide expectation in Central America that those changes can be very helpful in reaching solutions for the current conflicts.

SUBREGIONAL EFFORTS FOR PEACE AND SECURITY

It is difficult to think of a collective self-defense system in Central America. The great differences among countries and the dominant presence of an external actor appear as the main obstacles, at least in the short term. Nevertheless, it is equally important to emphasize the common interests of all Central American governments, despite their differences, to end war, conflict, and foreign intervention in the area.

A joint effort from El Salvador, Guatemala, Honduras, and Costa Rica to engage in an offensive war against Nicaragua is very unlikely today. The failure to reactivate the CONDECA (Central American Council of Defense) in the early 1980s had structural reasons. The Salvadorean army has only the capacity to focus on the internal conflict. To a certain extent that is also the case with Guatemala, in addition to its clear intention to keep its distance and not get involved directly in conflicts beyond its own borders.

The Honduran army, despite some modern weapons acquired in its military buildup, seems unable to succeed in an offensive war against Nicaragua without significant support and indirect intervention of the United States. Costa Rica does not have a formal army and has pretended a military and political neutrality in the subregion, being partially successful in both fields.

It is even more important that military confrontation among countries is very unlikely even with external support. The consequences of a decade of conflict and tensions have led most governments to search for a more stable and peaceful Central America. The populations of these countries are paying the highest price for the internationalization of internal and subregional conflicts.

The creation of the Esquipulas scheme in August 1987 showed the real concern of Central American governments to find solutions among themselves, without external interventions. They felt that Contadora had done its work and that it was time for the subregional actors to lead the process. The last meeting in Tela, Honduras, where demobilization of the Contras was approved by the five presidents, including Azcona Hoyo from Honduras, clearly indicates an increasing interest to move toward that direction.

CONCLUSIONS

Insecurity has become the lifestyle in Central America due to internal conflicts, as in El Salvador or Guatemala; to the threat of external aggression such as in Nicaragua and more recently Panama; or because of the presence of external actors as is the case of Honduras and to a lesser extent Costa Rica.

As noted earlier, Central America is only a close-up of the different features and variables that are involved in the concepts of security and strategy of the Latin American countries. Some preliminary conclusions can be drawn from this short review.

1. For most Latin American countries, security and defensive strategies are mainly concerned with the defense of independence and sovereignty, without significant threats among them, but in several cases internal conflicts and external intervention in those conflicts have greatly complicated the concepts of defense and national security.

2. Economically deteriorated situations and drug trafficking have become two main variables of national security in the region. Nowadays debt problems and insufficient growth have been increasingly related to the issue of sovereignty. Drug production and trafficking have, to a greater

extent, touched the concepts and situations of sovereignty, national security, and the missions of the military, creating new problems and scenarios. Drug trafficking, because of its internal and external implications, has become a major challenge to the power and control of some states.

3. The military forces in Latin America have tended to play a lower profile in politics in the last decade, although democracy is not strong enough to guarantee that civilians will be in power within stable political systems for the long run.

4. The United States has developed an active policy toward the region, according to its main interests: Communist containment, especially in Central America; a drug trafficking combat policy toward the countries where production is significant, particularly in Colombia, Peru, Bolivia, and Ecuador, and, with some differences in policy, Mexico; and with an ambivalent interest and concern for the economic crisis and the debt problem, although American policies have not changed the situation very much.

5. The hemispheric or inter-American system has suffered increasing deterioriation over the past decade. The United States has lost much of its moral and political leadership in the region. The OAS has become inefficient in dealing with political problems and disputes, especially between Latin American countries and the United States.

6. Central America remains the main focus of conflict in the region with no signs of a stable and peaceful situation in the short term. Moreover, the current strong differences between Panama and the United States can very easily worsen the subregional scenarios.

Facing these realities, the main challenges of security in the region can be summarized as follows:

1. The need for economic recovery to strengthen democratic systems and provide more stable political and social scenarios in most countries of the region.

2. Strengthening of the states' power and increasing international cooperation to be able to deal rapidly and properly with drug production and trafficking.

3. Political and operative recomposition of the national security concept in many of these countries, considering the economic crisis, the role of the military institutions to face current challenges, such as strengthening democracy, fighting the drug trafficking problem without becoming part of it, and maintaining defensive strategies vis a vis neighboring countries.

4. One of the main challenges becomes the transformation of the inter-

American system. The terms of dialogue and understanding between the United States and the region as a whole are at a very low level. The unilateral policies of the first toward the region, especially in Central America, have resulted in a difficult coexistence. Most Latin American countries disagree with American policy in the area—in the creation of the Group of Eight, we find the best evidence—and the U.S. government does not seem to show a real concern for creating a better framework for coexistence, which would give more attention to Latin American claims. Central America seems to be the core of the inter-American challenge for the immediate future.

For the purposes of a comparative analysis, all facts seem to indicate that defensive and strategic concepts in Latin America have little to do with the nonoffensive defensive concepts developed in Europe in recent years. Defensive strategies in Latin America are very different in nature because of the levels of conflict (Latin America is not the focus of the East-West conflict); arms involved (nonexistence of nuclear weapons); asymmetries among the different actors involved (we cannot talk of a strategic balance in the region); and interrelations of subregional scenarios with global conflicts (one dominant actor and other problems of security such as economic crisis, drug production and trafficking, and even territorial disputes) that are not of major concern in Europe. All these different features give Central America rather different scenarios.

Nevertheless, it is also a fact that there exists an interrelation between the different strategic concepts all around the world. What happens in Europe affects Latin America because the advances toward defensive concepts instead of offensive strategies have been possible also as a consequence of the improvement in relations between the two superpowers. As an example, we can now talk of emerging expectations among Central American actors to reach a better outcome in the subregion due to the decrease in tensions between the superpowers. In other words, a better understanding between the main actors involved in the East-West conflict will lower the significance of the conflict in Central America and give way to a more feasible bargaining framework for peace.

Part V

Defense, Disarmament, and the International Community

Chapter 10

Transition to Defense-oriented Configurations

Y. E. Fyodorov

The transition to defensive strategies and configurations of military potentials corresponding to them is shaped in the USSR by the restructuring of military forces on the principles of defensive sufficiency. The realization of this transition is one of the main and necessary conditions of ensuring lasting international stability and security in military-political dimensions. This process must acquire global scale and envelop both the central elements of military confrontation in the system of international relations and multiple regional focal points of military and military-political confrontation.

Generally the essence of transition to defensive sufficiency in the military field is more or less clear. It can be stated that the quantity, structure, qualitative characteristics, locations, and other parameters of armed forces and armaments, as well as military development and military activity in general, are oriented exclusively toward defensive purposes. This does not provoke suspicions from the other side and does not generate double meanings. Quantitative and qualitative characteristics of the military power and dynamics of its development should not create impulses that make the other side raise its military potential.

These general principles can be realized in different ways in applying them to different components of military power. For example, in the area of strategic offensive armaments, the defensive sufficiency apparently can be interpreted as guaranteeing the ability to deliver a second strike coupled with the inability to deliver a disarmament strike. In the field of conventional forces, the transition to defensive configurations must take away the possibility of using them in unexpected large-scale offensive operations.

The transition to the military defensive structure can be effected by unilateral actions and by the realization of measures aimed at establishing ceilings (probably zero options as well) on different components of military power, including particular types and systems of weapons. Such ceilings in many cases must be correlated with certain geographic or geopolitical zones. It is very important to stress that though unilateral actions are of significant importance, the basic way of transition to the defensive configuration of military forces is the development of and the enactment of multilateral agreements that fix mutual obligations on the parties and regulate steps to their fulfillment.

The process of transition to defensive military power structures determines to a great extent both general and concrete aims of arms control, restrictions, reductions, and elimination of different components of state military power. In the process of transition to defensive configurations of military power, the role and significance of international verification is by and large similar to the process of disarmament.

As experience in world policy shows, particularly experience accumulated in the areas of limitation and reduction of armaments, without reliable and adequate verification measures concerning agreements and unilateral actions, the process of rebuilding military forces to defensive sufficiency is impossible.

It is important to note that the effective control and verification that can guarantee adequate information about the armed forces and military programs of opposing sides can also reduce or eliminate the concerns about intentions and military potentials and can help to strengthen confidence in relations between countries. Experience also shows that as the share and significance of international and mutual forms of control and verification increase and have the potential to increase, the segments of military and connected civilian activities that are subject to control and verification also widen and will continue to do so.

The problem of control and verification has both technical and political aspects. As correctly noted by Soviet scientists, "Any agreement in the area of disarmament and the limitation of military activity can be provided with an adequate set of mutually complementing methods and systems of verification. The only remaining question is how far many states are prepared to go for the sake of a given agreement in terms of revealing delicate additional information, accepting constraints of their activities in related military and civilian fields not covered by the treaty, as well as whether, and to what extent, states will agree of their own free will to sacrifice their freedom of action in case of unforeseen circumstances."[1]

The transition to the defensive structure of military power will, it seems, stimulate two mutually complementing trends in the development of international verification systems. The first consists of increasing the quantity of different, sometimes unique, specialized verification systems of reduction or limitation of different components of military potential. The second is the creation of a limited number of international institutions, integrating these specialized systems in one way or another, fulfilling mutual functions, and accepting responsibility to a great extent for solving the tasks of verification.

The first trend is connected with the fact that the limitation, reduction, and elimination of components of military power during the transition to a defensive doctrine will necessitate solving a great number of problems, each with different distinguishing features and often unique character. This is determined by specific features both of the particular component of military power and the geopolitical zone in which corresponding measures are effected. As the transition to defensive strategies acquires greater depth and scale, the specialized, quite different and diverse functional and structural aspects of verification systems will grow.

I believe that the share of institutionalized verification systems will also grow, based on strictly regulated rules and procedures. This is determined to a great extent by the fact that the transition to defensive structures would inevitably cover a growing number of states that differ greatly in areas of military potential but are striving to retain their security by the process of verification. Such countries will wish to participate on equal terms in the process of decision making concerning all aspects of verification. Depending on the special features of arms limitation and reduction agreements, the systems of verification will acquire both global and regional character.

At the initial stages of transition to defensive structures that are taking shape today, a trend toward establishing independent specialized systems and mechanisms of control and verification would not by itself slow this process. However, the need is increasing to create structures that carry in themselves some elements of integration, eliminating duplication and execution functions common to all verification mechanisms. Above all, this is determined by the need to save the material and financial resources necessary to ensure the effective functioning of the verification process.

However, some problems now exist whose solution would seriously assist the transition to defensive structures of military power. Such a solution would be effected by the help of an international mechanism with a universal composition. Such a mechanism would not be a substitute for the specialized systems of international control and verification, but it could fulfill the following basic functions:

- Systematic collection and accumulation of information on military activities, forces, production, and such. In this connection, it is important to note the French proposal to create the International Agency for Space Control and the proposal by Finland to create the UN Bank of Data.
- Technical and expert services to states to provide effective control and verification of bilateral and multilateral agreement compliance, which should be developed during the process of transition to the defensive structures of military forces.
- Scientific and political expertise and analysis of technical and other rel-

evant aspects and problems of international control and verification, and preparation of recommendations on verification procedures and mechanisms to be included into any now and future agreements on the limitation and reduction of armaments.
- Coordination of the activities of separate specialized systems and mechanisms of international control and verification, including the elimination of duplication.
- Independent realization of the verification tasks of observance concerning the agreements in force or possible agreements on limitation and reduction of armaments with the consent of member states in cases when this is possible and useful from political and technical points of view.

The creation of an international institution with the above functions would mean a drastic reduction in the time spent in preparing treaties and agreements aimed at the transition to defensive structures. Ideally, one can imagine, for example, that such an international mechanism might develop different technical, organizational, judicial, and institutional schemes and models for verification of agreements on the limitation of components of military power in different geopolitical zones. In cases when states—if so interested—can come to an agreement, they can use the basic model of system verification that has been already developed.

It is important to underline that some concrete projects of such mechanisms do exist today. Besides the already mentioned proposals of France and Finland, there are proposals by Bulgaria, the USSR, and Czechoslovakia to discuss the question of creating, under the aegis of the UN, the mechanism of international control and verification of compliance with agreements on reduction of international tensions, limitation of armaments, and the military in zones of conflict (document AI-15/AC.I/15 dated June 13, 1988). Of special interest is the proposal of the "group of six" on creating the mechanism of international verification of nuclear tests.

Principally important is the connection of the international mechanism of verification that has a universal character with the UN system. It is due, first of all, to objectively raising the role of the UN in securing lasting international security. This organization has unique experience in solving international problems on the basis of a balance of interests. Even more, the UN has experience in effecting missions that deal closely with questions of international control and verification (such as the verification of compliance with the ceasefire agreements in some conflict situations). Besides, the UN Security Council has quite wide authority that probably can be used in cases where there is suspicion that some member state to an agreement is about to break or circumvent the agreement with the help of a state that is not a party to it.

Today, and probably in the near future the transition to the defensive structure of the military forces will include basically the central elements of the military confrontation, which are realized in NATO-WTO relations, (Here I mean first of all negotiations on conventional forces in Europe.) However, strong international security is impossible without globalization of this process, spreading it to the zones of regional conflict in Asia, Africa, and Latin America. But restructuring the military forces of states participating in such conflicts on defensive principles will inevitably face huge difficulties. The sources of these difficulties rest in the fact that in the Third World military conflicts are, and unfortunately can remain, the possible way of solving both internal and international contradictions. Under such conditions, efforts toward limitation of the scope and blocking mechanisms of escalation of regional military conflicts are necessary. Very important trends for such efforts are the strengthening of nonproliferation of nuclear weapons, and prevention of the spread of chemical weapons and ballistic missiles technologies with the corresponding mechanisms of verification. For example, it is necessary to develop further the MAGATE safeguard on the activities of states in the field of nuclear energy, and to improve the fact-finding mechanism undertaken by the UN measures of strengthening the Geneva Protocol of 1925 on the prohibition of chemical weapons.

Corresponding procedures can be developed for other multilateral treaties and agreements, including those existing today, for example, the convention of 1981 on the prohibition or limitation of using particular systems of conventional weapons that can be considered as producing excessive harm or having an unwanted effect.

In conclusion, I stress that openness in the military-political sphere and, hence, creation of the mechanisms that permit realization of the transition are now extremely important. The restructuring of military potentials on the principles of defensive sufficiency cannot be effected without realistic evaluation of the level of external threat. Such an evaluation often has a significant subjective component. In the absence of adequate knowledge of the military strength of a potential adversary, states are inclined to base their military planning on the "worse case scenario," and that, in turn, will lead away from defensive sufficiency and to the creation of offensive military potential or its elements.

All this produces new requirements for the technological, judicial, and institutional sides of international control and verification, which become both the condition and means of greater openness.

Notes to Chapter 10

1. "Disarmament and Security, 1987," *IMEMO Yearbook,* (1988), p. 311.

Chapter 11

Defense, Disarmament, and Collective Security

Aga Shahi

The Vienna talks between the 23 members of NATO and the Warsaw Pact hold the prospect of an agreement that eliminates military disparities between the two sides and strikes a stable and secure balance of conventional forces in Europe at lower levels. The idea that conventional armaments in a region where two great military alliances confront each other with enormous concentrations of weapons and armed forces and where a conventional conflict could escalate to the nuclear level should be regulated on the basis of defensive principles. It should rule out suprise attack or large-scale offensive action. These ideas have led to a debate in military and political circles on the concept of "nonoffensive" or "nonprovocative defense" in the Central European theater or the European continent as a whole (the Atlantic to the Urals).

A leading theorist, Anders Boserup, has succinctly defined the concept of nonoffensive defense as a bilateral relationship in which "both sides have forces that are stronger than those of the opponent when used in attack"—that is, a situation in which both sides have "ample forces for defence and, by implication, insufficient forces for attack."[1]

Proponents have argued that efforts to transform the global offensive capability of the major powers into a posture of nonoffensive defense could best be pursued as part of a larger political and economic restructuring of the international system. It has been contended that "Any application of nonoffensive defence to global force projection must be part of a much wider reappraisal of foreign policy extending to North-South relations, trade, development assistance, and security of resources."[2] Promoting nonoffensive defense in this area "would require the most research, integrating the work of defence, foreign policy, and development analysts."[3]

The strategic balance between the superpowers, which is based on nuclear arsenals, is a highly important area where the possibilities of nonoffensive defense need to be explored. A specific issue in this regard is the question of first use—the doctrine of extended deterrence—which continues to form an essential feature of NATO's military strategy toward the WTO. By virtue of the policy of extended deterrence, nuclear weapons have acquired an offensive posture. It has been aptly remarked that "A strategy envisaging first use of nuclear weapons may have made a kind of perverse sense during the years of Western nuclear superiority. But with nuclear parity it makes no sense at all, except as a temporary expedient until an acceptable defence concept can be put into effect."[4]

A no-first-use stricture would be a step toward giving nuclear weapons a nonoffensive character. But more changes would be necessary to achieve nonoffensive defense in the nuclear field. In addition to substantial reductions in strategic nuclear weapons, as envisaged in the on-going START negotiations, it would be necessary to restructure the nuclear deterrent, as well as contain the technological arms race, so that threat perceptions can be effectively addressed. As has been pointed out, "even a START agreement that involved deep cuts, which may look attractive, could simply dismantle obsolete systems and even encourage the shift towards counterforce weapons."[5] Several measures have been suggested to initiate the process of reducing the offensive character of the nuclear forces of the superpowers. One of these has been the long-standing issue of a comprehensive test ban treaty. Another proposal that also aims at limiting the qualitative development of nuclear weapons is a prohibition on nuclear forces modernization.

The concept of nonoffensive defense should also be examined in the context of reducing the global interventionist capability of the major powers. Some advocates of nonoffensive defense have drawn attention to this area, though the subject has yet to be examined or discussed in any depth. The bitter experiences of the superpowers with their interventions in Vietnam and Afghanistan may not necessarily make them less likely to use their interventionist capabilities in the future. A U.S. study—"Discriminate Deterrence"—projects a highly unilateralist interventionist role for the future. Were they to opt for nonintervention, the door would be opened "to a global defense-oriented restructuring of US and Soviet conventional forces."[6]

It has also been argued that transforming the global capabilities of the superpowers would enable the concept of nonoffensive defense to be more fully realizable in Europe. An example cited in this connection is that "most if not all of the problems of implementing a full-scale non-offensive defence in Europe, that is, abolishing all long-range attack systems and all nuclear weapons except those of the French and British, would disappear if there were a ban on such systems throughout the U.S. and Soviet armed forces."[7] In other words, nonoffensive defense in the global arena would not only help to eliminate superpower interventions in the Third World but would also improve conditions for a fuller application of the concept in Europe.

The verification requirements for agreements on nonoffensive defense would vary according to the scope of arrangements. Agreements could range from limited nonoffensive defense zones along the border to arrangements affecting the size, weapons, training, logistics, doctrine, war games, and maneuvers of the armed forces in a manner that they are seen in their totality to be capable of a credible defense without any reliance on the use of nuclear weapons, yet incapable of offense.[8] Verification measures for such agreements would necessarily have to be of an extensive and intrusive nature.

NONOFFENSIVE DEFENSE AND REGIONAL SITUATIONS

Developments taking place in Europe on both sides of the divide are likely to enhance the status of nonoffensive defense. But there is much that has to be done to focus attention on the concept's relevance to other regions. The existing research related to Europe has produced a number of unilateral and reciprocal models, some of which may be of general interest. But, on the whole, the various approaches may not be readily applicable elsewhere. A major problem in relating proposals for Europe to other regions is that, whereas there is a clear-cut division of Europe into two alliance systems and the principle of parity operates in East-West arms control negotiations, a more complicated situation exists outside the continent. The principle of parity would be difficult to introduce in other regions because of considerable variations in the size and military strength of states. Other difficulties include the extraregional dimension—that is, the role of global powers as well as security linkages between regions. Interstate conflicts, territorial disputes, and ambitions for regional dominance, quite apart from tensions arising out of military imbalances, would also pose impediments.

The major arms suppliers could help to limit the offensive capabilities of recipient countries through reciprocal or even unilateral controls over the transfer of offensive weapon systems and military technologies. In the past, such efforts proved singularly unsuccessful because relations between the superpowers were confrontational and bedeviled by acute rivalry in important regions. Arms supplies to strategic allies loom large in their foreign policies. The United States and the Soviet Union have been the leading arms suppliers. In 1988, for example, they accounted for more than $9 billion each. Between them, they are responsible for nearly two-thirds of all weapons sales to the developing countries with the Middle East absorbing two-thirds of all the weapons delivered. Arms transfers to Iraq and Iran, governmental as well as clandestine, prolonged the Gulf war to eight years, cost hundreds of thousands of lives, and inflicted damage estimated at more than $400 billion.

The continued improvement of East-West and U.S.-USSR relations could create a more favorable environment for reaching agreement to exercise restraint in the supply of offensive weapon systems, limiting them to legitimate security requirements. However, reducing military capability to nonoffensive defense would require more than the willingness of the major arms suppliers to exercise restraint. It would also call for willingness on the part of recipient countries to agree on modalities for moving reciprocally toward a nonprovocative defense posture. Furthermore, the growth of clandestine arms traffic with possible links to the drug trade could prove difficult to control.

It is clear that nonoffensive defense would have to contend with a more challenging situation in regions outside of Europe. Systematic research and informed discussion would help to clarify the situation. Research effort needs to be directed toward exploring possible approaches for strengthening regional and international security through the reduction of offensive conventional weapons or more modest steps such as the introduction of confidence-building measures (CBMs).

SECURITY-BUILDING MEASURES

Confidence-and security-building measures (CSBMs), though not a substitute for substantive arms control agreements, can advance the cause of security and be applied extensively. Although CSBMs have been pursued in Europe for over a decade and tangible military progress has been made, there have been relatively little effort and much less progress in other regions where CSBMs could prove extremely useful as well as serve as a starting point for arms control. If models of nonoffensive defense proposed for Europe are not easily applicable elsewhere, the experience with CSBMs could prove valuable. They can help to reduce the threat of surprise attack and also inhibit the threat of force for political intimidation.

There are several CSBM arrangements, some more limited than others, in Third World regions. A major disengagement regime has been in force between Israel and Egypt initially under the 1975 Sinai agreement and later under the 1979 peace treaty. The UN had monitoring responsibilities under the Sinai agreement, but since 1979, the United States has assumed the responsibility of verifying compliance. Since 1974, a UN peacekeeping force has also been implementing a disengagement zone in the Golan Heights separating Israeli and Syrian forces. In Sinai, the area included a buffer zone separating Israeli and Egyptian troops, which was entirely monitored by the

United Nations Emergency Force (UNEF). The UN's role virtually ended with the signing of the 1979 treaty between Israel and Egypt. The UN has, however, continued to monitor the disengagement zone in the Golan Heights while maintaining a small supervisory presence in Sinai—the UN Truce Supervisory Organization (UNTSO).

A more limited confidence-building measure has more recently been in force between India and Pakistan. In December 1988 the two countries signed an agreement that forbids attacks on each other's nuclear facilities.

In Latin America, efforts have been made by Brazil and Argentina to reduce each other's threat perceptions regarding the danger of nuclear proliferation. Initiatives to resolve the crisis in Central America include CBMs aimed at restoring peace and stability in the subregion. These approaches go far beyond the essentially military framework of the CSBM measures applied in Europe, to address political, economic, and social problems in the area. The Contadora Act of September 1984, signed by Costa Rica, El Salvador, Guatemala, Honduras, and Nicaragua, was aimed at establishing a "balance of forces in the region, thereby precluding any country from having the military capability to establish its hegemony." The act envisaged a series of mechanisms to increase confidence among the five states on security-related issues. Among these were advance notification of military maneuvers along frontiers, invitations to foreign observers, elimination of intra- and extraregional arms traffic that could be directed to destabilization of governments, and prohibition against the introduction of new weapons systems.

The institution of a variety of CBMs in South America, can help to rebuild a general environment of trust that must precede lasting solutions to the underlying causes of conflicts between the nation's of this continent. Some of the successful CBM examples are the mutually agreed-upon maritime boundaries in the Beagle Channel between Argentina and Chile, and the unilateral cuts in Argentina's military expenditures by the Alfonsin government. British-Argentine relations are critical for re-establishing a regional atmosphere of trust as are Brazilian-Argentine relations. In regard to the former, CBMs, as for example, phasing out the 150-mile protection zone around the Falklands/Malvinas, followed by a reciprocal gesture, could start a process toward breaking the British-Argentine impasse over the status of these islands. Should the deadlock persist, pressure within Argentina to exercise its nuclear weapons option and to build nuclear submarines could greatly increase.

In Africa, tensions and armed conflict between and within states in the postcolonial era have lent importance and urgency to the implementation of confidence-building measures. The earliest instance of CBM was the UN peacekeeping operation in the Congo (Zaire) in 1961. One of the most farsighted steps taken by the Organization of African Unity (OAU) was its solemn declaration in July 1964 that all member states pledge themselves to respect the borders as existing on achievement of national independence. The declaration has pre-empted many claims and disputes over boundaries that otherwise could have disrupted the territorial integrity of the newly independent African states.

Other causes of conflict—territorial disputes, insurrectionary movements, ethnic strife—however, have led to wars and internal conflicts. The Horn of Africa, Western Sahara, West, Central, and Southern Africa have been theaters of violence and bloodshed at one time or another. Hence the need exists for military and security-related confidence-building measures to reduce tensions or misunderstandings in these subregions.

The most serious threat to sub-Saharan Africa's security, however, stems from South Africa, whose neighbors have been the victims of repeated acts of aggression, subversion, and destabilization by the apartheid regime. The continuing threat from South Africa is perceived by other African states as compounded by Pretoria's nuclear weapons capability. The prospect, however, for the introduction of CBMs between the apartheid regime and its neighbors is by no means encouraging.

In other subregions, the prevalence of inter- and intrastate conflicts remains a compelling reason for the application of CBMs. The Economic Community of West African States (ECOWAS), consisting of 16 nations, aims to establish not only a common market and promote economic development, but has emerged as an embryonic subregional system of collective security. The 1978 Protocol of Nonaggression has been conceived as insurance against external aggression and externally supported insurrection against any member of the community. A Defense Protocol concluded in 1981 provides for mutual assistance in emergency situations.

The OAU has revealed its potential for a peacekeeping role by its initiative at the end of 1981 to deploy a force of 3,000 troops in Chad. Though the operation was not a success because financial resources and logistic support could not be mobilized on a pan-African basis, the OAU demonstrated its ability to raise an African peace force to deal with an African conflict.

In southern Africa, six frontline states and three others in geographical proximity to South Africa have established the Southern African Development Coordination Conference (SADCC) to promote economic cooperation in the region and to minimize their dependence on South Africa. They have also committed themselves to a code of conduct that calls for "avoidance of territorial expansionism and occupation of each other's territory and refusal by each country to serve as a base for subversion and destabilization by another member country." A formal defense and security provision to counter South African aggression is not yet contemplated by the SADCC.

In Asia, the North Pacific, Southeast Asia, and the Persian Gulf are three subregions where CBMs could help to reduce tensions and promote a greater sense of security. In his Vladivostok speech of July 1986, Mikhail Gorbachev put forward a proposal for CBMs for the Asia-Pacific region on the model evolved through the Helsinki process. But he soon conceded the problems entailed in their application because of the geopolitical differences between this vast region and Europe. The absence of diplomatic relations between some of the key countries to be included and the positions adopted by the nuclear-armed navies of the two superpowers in the vast expanse of the Pacific Ocean are major impediments to the introduction of CBMs.

A more limited application confined to the land forces of the littoral and hinterland states of the North Pacific would perhaps encounter less intractable difficulties. One suggestion is for a Consultative International Commission on the North Pacific, which would include Japan, the two Koreas, China, Taiwan, Mongolia, and the far East part of the Soviet Union, plus the United States and Canada, in order to circumvent the problems of negotiating CBMs in the absence of mutual recognition on the part of some of the states of the subregion.

CBMs need to be introduced in the North Pacific in the near future, because proliferation of nuclear weapons in the Korean peninsula, already one of the most heavily fortified and dangerous areas of the world, is a distinct possibility.

The conflict between China and Vietnam over the Spratley Islands and the tension along their common border, as well as the claim to the islands by some of the other area states, marks this part of Southeast Asia as one where the introduction of CBMs is necessary to prevent an armed clash. Until the Cambodian crisis is resolved, however, and Sino-Vietnamese relations improve, any consideration of such measures would appear to be premature.

In the Persian Gulf, foreign naval vessels continue to patrol the sea lanes. The cease fire between Iran and Iraq remains tenuous. Until a peace treaty is signed and ratified by the two sides, the security situation in the Gulf cannot be stabilized. The waterway is of major importance to the industrialized as well as developing countries. The dangers that will arise from the projection of power in its confined but vital waters call for a security regime that insulates it from the naval presence of all but its littoral states. For their part, the littoral and hinterland states of the Gulf must demonstrate their willingness to agree to naval and land-based CBMs to guarantee freedom of navigation in this arm of the Indian Ocean.

South Asia is one of the regions where CSBMs could play an important role toward alleviating threat perceptions that could lead to conflict or a more competitive and expensive arms buildup. Military exercises have posed problems. An example is the dangerous spiraling of tension a few years back when India carried out its massive Operation Brasstacks along Pakistan's southern border of Sind. An agreement on CSBMs between the two sides would increase trust and confidence, making the intentions and actions of each clearer and more predictable to the other.

The stationing of UN observers on one's own side of the border is a confidence-building measure that could inhibit an attack by its neighbor.

The 1986 Stockholm agreement on CSBMs between NATO and WTO provides for more wide-ranging measures and stricter verification provisions than the preceding Helsinki agreement. Significantly, the Stockholm agreement covers the whole of Europe unlike that in Helsinki, which did not fully extend to the European part of the USSR. It provides for intrusive verification measures such as compulsory on-site inspection by land or air or both. Prior notification is obligatory for activities involving 13,000 or more troops and 300 or more battle tanks. Air force exercises involving 200 or more sorties are also liable to advance notification by more than a month. Activities with 17,000 or more troops are subject not only to prior notification but also to direct observation. More significantly, the Stockholm agreement restricts large-scale exercises. Those involving between 40,000 to 75,000 troops require one year's advance notification. A two-year prior notification is laid down for exercises exceeding 75,000 troops. These measures have considerably reduced the risk of surprise attack and military confrontation in Europe.

The European experience cannot, of course, be automatically transferred to other regions. Outside Europe, CSBMs could be initiated on an incremental basis, as has been the case in that continent, and also in a manner that takes account of the specific characteristics of the various regions and other problems of a structural nature. For example, the zone of application could be kept limited, but substantial restrictions as well as verification measures could be imposed within that zone. The latter would partly compensate for the former. CSBMs should also address the nature of threat perceptions. For example, if the perception of threat is not simply land based but also air or sea based, these aspects cannot be overlooked if CSBMs are to make a difference to the security situation. Similarly, restrictions could be made more strict to aspects that are particularly threatening.

NAVAL CBMS—ZONES OF PEACE AND NUCLEAR-FREE ZONES

Naval CSBMs would need to be considered in the context of converting the Indian and other oceanic regions into zones of peace. The concept of Zone of Peace in various proposals implies a nuclear weapon-free status, or, as a minimum, nonnuclear status for the zone countries.

The Sri Lankan proposal for declaring the Indian Ocean as a Zone of Peace was endorsed by the UN General Assembly in 1971. An ad hoc committee on the Indian Ocean was subsequently established. In 1979, 44 Asian and African littoral and hinterland states agreed on a number of principles for the zone and called for a full-scale conference to establish it. The Western powers, however, do not agree to the elimination of foreign military bases and their military presence or to the denuclearization of the Indian Ocean. The absence of consensus within the committee continues to cloud the prospects of holding the proposed conference, which has been postponed for the seventh time. It is now proposed during 1990.

Conditions in the Indian Ocean have changed since the proposal was first made and even since the time that agreement on the principles for creating the zone was reached. The main thrust of the Zone of Peace proposal remains halting the escalation and eliminating the military presence of the superpowers "conceived in the context of great power rivalry." The Soviet naval presence peaked out in 1980 and has decreased since then. Although the United States has upgraded its base facilities in Diego Garcia, its capabilities under the Central Command (CENTCOM), established in 1983 and augmented substantially by its Joint Task Force Middle East (JTFME) since that time, are directed at the Gulf and apparently are not a direct manifestation of "great power rivalry."

The concept of the Indian Ocean as a Zone of Peace can no longer be pursued on the restrictive basis of addressing the military presence of the superpowers alone. Such an approach would not only limit the scope for strengthening peace and security in this vast and most strategic expanse of water but would ignore the altered situation that is likely to result from intraregional naval and political developments and their linkages with great power military presence and rivalry. One such likely development is the proposed cooperation between a superpower and a regional power in the form of joint naval exercises in the Indian Ocean. A second is the announced readiness of Singapore to grant naval base facilities to the United States. Singapore considers the U.S. military presence as essential for the maintenance of a balance of power in the Indian Ocean. A third likely development is intraregional naval competition as a result of the rising naval power of a regional state and its power projection capability from the Strait of Malacca to Aden, including the Gulf. Australia, Indonesia, and possibly Iran and some members of the Gulf Cooperation Council seem sufficiently aware of the destabilizing situation that may consequently arise, as to augment their own naval forces. Acquisition of nuclear submarines by any of the regional states would extend its nuclear presence into new areas.

The Indian Ocean conference, whenever it is held, must inevitably confront the new realities. The presence of nuclear weapons in the region and the incipient intraregional naval arms race as well as the linkages between the regional states and the superpowers would need to be addressed. CBMs alone could hardly provide assurance of security. What is called for is an oceanic regime that would regulate naval competition and provide for the nonuse of force, nonintervention, and noninterference, for peaceful settlement of disputes and respect for sovereign equality and renunciation of domination or hegemony. Guarantee of freedom of navigation and nonuse of nuclear weapons against the nonnuclear littoral and hinterland states must also figure among the objectives of a conference on the Indian Ocean.

As an interim arrangement, some confidence-building measures proposed by the Soviet Union—apparently in the context of its new defensive doctrine and its concept of comprehensive security—could be examined for implementation. The Soviet CBM proposals include restrictions on the number and area of annual large-scale naval and air exercises in the Indian Ocean, as well as timely notification and observation of such maneuvers. A second Soviet CBM calls for banning all exercises and war games in the international straits and adjoining areas. In terms of substantive arms control measures, a significant Soviet proposal is for an international agreement that guarantees the safety of the sea lanes in the Indian Ocean, including the Gulf and also the straits of Hormuz and Malacca. The idea of establishing a UN naval force for maintaining freedom of navigation was suggested by the Soviets a few months before the ceasefire between Iran and Iraq. Another Eastern European CBM proposal is for the prevention of military incidents at sea.

Besides confronting these issues, the Zone of Peace proposal must take into account other relevant aspects of the security situation as it has developed. This would include the contribution that the zonal states should make toward the elimination of tension in the Indian Ocean, for example, by observing military restraint and reducing their naval forces to lower levels.

The establishment of the Indian Ocean Zone of Peace could be facilitated by resumption of bilateral talks between the superpowers to limit their presence in the zone. These talks began in June 1977 but were suspended by the United States in early 1978, ostensibly because of Soviet military assistance to Ethiopia against Somalia and abandoned after the Soviet military intervention in Afghanistan at the end of 1979. The talks, however, did apparently result in a draft agreement that included limits on the size and frequency of their naval forces deployed in the Indian Ocean, as well as a freeze on permanent forces and military facilities.

Since 1986, 21 states of Africa and South America have joined in establishing a Zone of Peace and Cooperation in the South Atlantic as a confidence-building measure. In their declaration, they called for states outside the region to reduce and eventually eliminate their military presence, the nonintroduction of nuclear weapons and other weapons of mass destruction, and the nonextension into the region of "rivalries and conflicts foreign to it." Effective implementation of the declaration would, of course, depend on the cooperation of the three Western nuclear weapon powers who take the position that the right of freedom of navigation in the South Atlantic cannot be denied to the nuclear components of their naval forces.

Peace and nuclear weapon-free zones represent a regional approach to the prevention of geographical proliferation of nuclear weapons. Since the Treaty of Tlatalelco of 1967, which established the Latin American and Caribbean nuclear weapon-free zone, only one other NFZ has been established—in the South Pacific under the Treaty of Raratonga. The zone extends from Western Australia into the South Pacific Ocean up to the western limits of the ocean area of the Treaty of Tlatalelco, and takes in, besides Australia and New Zealand, the island states of the South Pacific. The parties to the Treaty of Raratonga have pledged themselves to keep the region free of nuclear weapons, but the Western powers insist on the right of free navigation for their nuclear armed warships through the zone.

Two regional initiatives, namely, nuclear weapon-free zones for South Asia and for the Middle East with a view to contributing to security in two of the most conflict-prone regions, have been unsuccessful. India in South Asia and Israel in the Middle East do not favor the proposals. Consequently, the prospect of transforming the two regions into nuclear weapon-free zones is not encouraging.

The outlook for the denuclearization of the continent of Africa is no better. The efforts of the OAU over 25 years to keep Africa free of nuclear weapons have foundered on the rock of South Africa's nuclear ambition. Now that the apartheid regime admits its nuclear weapons capability, the promise of Africa's emergence as a continental NFZ is not likely to be fulfilled. Should any black African state choose to follow the nuclear path—40 out of the 50 African states are parties to the Non Proliferation Treaty (NPT)—confidence and trust among the regional states would be further eroded and African security seriously undermined.

Peace and nuclear-free zones represent regional approaches to the prevention of the geographical spread of nuclear weapons. Although there is extensive scope for the introduction of confidence-building measures in various regions of the Third World, to prevent surprise attack or the launching of large-scale operations, these measures have been applied in only a few of them. Peacekeeping forces have been interposed to separate combatant sides in some areas of conflict, but they are not an instrument of war prevention. Outside of Europe there have been no regional initiatives for arms control or conventional disarmament, nor for the reduction of military expenditures, which absorb large amounts of scarce resources that could be better used for development and social welfare purposes. The emergence of regional nuclear weapon powers and the danger of proliferation of chemical weapons compound the threat to security in many regions outside of Europe.

Almost all the regional organizations that exist have not been designed or structured for purposes of war prevention or conflict resolution. It is only a global organization, such as the United Nations, that can hold the promise of collective security to states outside the military alliances of the two superpowers.

COLLECTIVE SECURITY

With the easing of Cold War tensions, the United States and the Soviet Union are much better placed today to improve the basic structure of international security by making the UN more consequential in matters falling under Chapter VII of the UN Charter. Current international political trends provide a unique opportunity to make a break with the past pattern of global politics. If it is still premature to expect the great powers to put together the universal collective security system envisaged in that Chapter VII of the UN Charter, a more limited arrangement could be pursued initially and tested out.

A beginning could be made by breathing life into some of the provisions of Chapter VII, in particular, Article 51, which recognizes the right of individual or collective self-defense. This article has been frequently invoked in justification of the establishment of the rival alliances of the two superpowers. But its scope is not limited to military alliances. The right of collective defense could be extended to arrangements outside of military pacts to deal with threats to the security of the nonaligned nations resulting from the proliferation of nuclear and chemical weapons. Negative as well as positive guarantees to prevent the use or threat of use of these mass destruction weapons against the nonaligned, non possessing states would promote a greater sense of security. To be considered credible, the guarantees must be provided jointly as well as individually by the Permanent Members of the Security Council.

Despite the 1925 Geneva Protocol, which prohibits chemical warfare, chemical weapons have been used in recent regional conflicts. At present more than 20 states are believed to be capable of manufacturing such weapons.

The negotiations in the 40-member Conference on Disarmament at a convention to prohibit the possession and use of chemical weapons have no doubt made remarkable progress. But the instrusive verification procedures to secure compliance with the convention, which were reported as agreed upon by the two superpowers, also must be acceptable to the rest of the membership. Furthermore, any assistance to be extended by the executive organ to a party that is a victim of chemical weapon use, or feels threatened by them, still must be formulated. The credibility of the assistance would depend upon its timeliness and efficacy. But whatever the powers and functions that may be assigned to this executive body, the Security Council must remain the primary organ of the last or even the first resort for member states of the UN in threat contingencies.

It is still uncertain if the Conference on Disarmament would lead to the conclusion and early implementation of a multilateral agreement on a comprehensive and total prohibition of the development, production, stockpiling, and use of chemical weapons and on their destruction, or only on the geographical nonproliferation of these weapons. Nor is universal adherence to the convention to be taken for granted, in particular by those nonnuclear states in uneasy coexistence with regional nuclear powers. The former could be expected to demand credible security assurances against the use or threat of use of chemical weapons as the price for renunciation of their chemical weapons option.

Equally urgent is the prevention of the use or threat of use of nuclear weapons against the nonnuclear states. Unlike the convention on chemical weapons under negotiation, there is no legal instrument that prohibits the use of nuclear weapons. The so-called positive assurances, written into Security Council Resolution 225 of 1968, of assistance by three of five Permanent Members to nonnuclear weapon states if threatened with the use of nuclear weapons fall short of the kind sought by the latter. The assistance promised is subject to the right of veto. Furthermore, the identical statements made by the three powers amount only to an "intention" to provide assistance. They fall short of a commitment.

As far as negative security assurances are concerned, the decade-old negotiations in the Conference on Disarmament have made no progress in removing the conditions, limitations, and exceptions contained in the different unilateral declarations of four of the Permanent Members not to use or threaten to use nuclear weapons against nonnuclear states. Only China has extended a guarantee of nonuse in unqualified terms. The conference has yet to evolve a common formula to be included in a binding international instrument that meets the concerns of the nonnuclear states.

The NPT may have slowed the horizontal proliferation of nuclear weapons, but it has failed to stem the spread. Since the signing of the treaty in 1968, three additional states have emerged as de facto nuclear weapon powers. The next decade could see the production of the atomic bomb by some countries that do not now possess it. Prevention cannot be ensured solely by embargoes on nuclear supplies or by discriminatory sanctions. Credible assurances of security would be more efficacious as a disincentive and an important step toward strengthening the nonproliferation regime.

The spread of chemical and nuclear weapons has become an even greater cause of concern because of the proliferation of medium-range ballistic and shorter range missiles and the technology of their manufacture. This destabilizing development has added a new dimension to the threat to non-aligned nations. Missiles would enable such weapons to be used for deep counterforce strikes, because of their greater range and accuracy, as well as retaliatory strikes against large civilian targets and vital economic assets not easily reachable otherwise. The Middle East and Southwest and South Asia are regions of particular concern, not simply because the proliferation of missiles and missile technology have been more pronounced in these areas, but also because of the endemic tensions and disputes that could lead to conflicts in which these highly offensive capabilities can be put to horrific use.

The basic assumption underlying the Chapter VII provisions of the UN Charter is that in situations of threat to international peace and security, the five Permanent Members of the Security Council would act unanimously. This assumption has proved false. The Cold War that broke out when the UN came into existence has for more than 40 years hamstrung the Security Council from exercising its special responsibility for the maintenance of peace. Now that the superpowers are moving away from confrontation and involvement in regional conflicts to cooperation, the world organization is, in Sir Brian Urquhart's words, at the starting line it was in 1945. "The prospects of realizing the dreams of 1945, as expressed in the UN Charter, are better than at any time in 40 years."[9]

President Gorbachev of the USSR has unfolded, in his address to the UN General Assembly in December 1988, his vision of comprehensive collective security that would embrace security against military as well as non-military threats. President Bush of the United States has spoken of "a new world within reach." The time has come for the two world leaders to give shape and substance to their grand designs that have much in common, in a manner that provides security not only to those nations that dwell in a common European house, but to all peoples who inhabit the common global village.

Notes to Chapter 11

1. Andres Boserup, "A way to undermine hostility," *Bulletin of the Atomic Scientist* (September 1988), p. 16.

2. Paul Rogers, "The nuclear connection," *Bulletin of the Atomic Scientist* (September 1988), p. 22.

3. *Ibid.*

4. Boserup, p. 18.

5. Rogers, p. 21.

6. Randall Forsberg, "Toward a non aggressive world," *Bulletin of the Atomic Scientist* (September 1988), p. 49.

7. *Ibid.*, p. 53.

8. Hopmann, P. Terrence and Frank Barnaby (ed.), *Rethinking the nuclear weapons dilemma in Europe*; foreword by Harlan Cleveland. (London: MacMillan, 1988), p. 137.

9. Brian E. Urquhart, *The United Nations system and the future*. (Royal Institute of International Affairs—United Kingdom), 65(2) Spring 1989, pp. 225–231.

Conclusion

Reflections on a Nonoffense Defense

Jefferey Laurenti

When defense policy thinkers from a dozen and a half diverse countries first get together to ponder how and where "nonprovocative defense" connects to reality, the opening hurdle consists of arguing out the meanings of the basic terms of reference. It is no surprise, then, that much of the symposium on nonoffensive defense strategies organized by the United Nations Association of the United States and the United Nations Institute for Disarmament Research was devoted to defining the essential characteristics of a concept that had heretofore been the subject of a highly region-specific debate. Simply by inquiring into the transferability of the idea—one still controversial even in the regional setting that gave it a name (or a collection of names)—the contours of the concept become more sharply defined.

It is an exercise reminiscent of the South Asian tale of 10 blind men trying to understand an elephant based on the small part of its body each is able to touch. The initial problem is to understand the seemingly elusive nature of "nonoffensive defense," or NOD. In the taxonomy of defense policy, is NOD a separate species, or is it simply a seasonal variant of an already familiar animal—confidence-building measures, balanced force reductions, or whatever? What kind of habitat is necessary for this creature to thrive? The argument is complicated by the fact that the animal has not actually been observed in its original terrain of Central Europe, only imagined (and thus easily derided by skeptics as a unicorn); yet its description in a multilateral setting yields persuasive accounts of its sighting in unexpected places, like the Sinai and Central America. Could NOD actually be a Third World hybrid awaiting Europe's discovery?

So what distinguishes the species? There is wide agreement that the notion of NOD partakes of some of the same qualities as other items on the standard menu of detente and disarmament. In the most probable scenarios it may indeed involve confidence-building measures, demilitarized zones, stringent verification, and transparency. Yet it is not clear that nonoffensive defense *must* involve these, even though these will almost certainly be indispensable ingredients of a NOD agreement where states or alliances of comparable military power embrace the doctrine on a reciprocal basis.

Defensive defense strategies can also be military options on a unilateral basis. This was in fact the context in which the strategic theory was first articulated in Western Europe early in the 1980s (even if, in that era sometimes viewed retrospectively as burdened by "stagnation," the theory was roundly rejected as unnecessarily risky and defeatist). Whereas unilateral recourse to nonoffensive defense is probably politically implausible where opposing sides are roughly at military parity, a one-sided NOD is often a smaller state's only real option for defense against large neighbors. Singapore's "poison fish" defense strategy is a classic NOD of the weak: making the prize militarily indigestible to an attacker, or at least making the price of successful attack so high that he is deterred from attempting it. Whether NOD is the unilateral defense policy of the weak or a mutually negotiated defense regime of the strong, one of its essential ingredients is reliance on *deterrence by denial*.

This simply means that the goal of any nonprovocative defense strategy is to deter war by ensuring that an adversary cannot conquer, rather than by threatening the adversary with substantial destruction of his own home territory (deterrence by punishment). There is, of course, always the risk that the attacker may be able to muster such overwhelming offensive force that his conquest is not denied. For small and weak states, necessity compels them to live with this risk. Large powers have the resource capabilities to add the perceived extra (if costly) margin of security they can buy through deterrence by threatened punishment. For these, recourse to nonoffensive defense—and their abandonment of deterrence by punishment—almost certainly requires a similar renunciation by their potential enemy. And it is this context—NOD by choice, rather than NOD by necessity—with which the policy debate is concerned.

The most essential element of this elective, reciprocal NOD is political. The parties involved must conclude that they can, and probably have to, live with each other (and can do so within existing borders), that war would pose excessive dangers and that the costs of preparedness against possible war are intolerably burdensome. Irredentism, the denial of an adversary's legitimacy, passionate claims over land or populations are all incompatible with the political prerequisite for nonprovocative defense; all imply that a cause in dispute is so important that the use or threat of force must remain an available option.

The other essential, and indeed defining, element of nonoffensive defense is military: military forces must be structured, deployed, and equipped with weapons in such a way that they cannot be easily projected deep into the adversary's territory. It is in this sense that NOD can serve as the ultimate confidence-building measure. As each side is satisfied that under a NOD regime its potential adversary cannot launch a surprise attack, the momentum grows for further rounds of arms reductions.

Additional criteria for successful reliance on NOD are often advanced, as they were at the symposium. Some, pointing to the North Atlantic Treaty Organization and the Warsaw Pact alliances as the model, argue that bipolarity is an essential ingredient; some, that contentious rivals must have roughly equal military capabilities; some, that strategic depth of territory (especially if it is mountainous or heavily forested) is a necessary guarantee of NOD military feasibility. But these characteristics of the Central European theater—where, after all, NOD remains an untested theory—seem more to be simply conducive to, rather than prerequisites for, successful implementation of a nonprovocative defense regime.

It is true that there are particular conditions that have brought nonoffensive defense concepts out of the closet and onto the East-West agenda. The Soviet Union needs some economic breathing space to enable *perestroika* to produce results domestically, and deficit-dogged Western governments and taxpayers are eager to hold down defense spending. But even these are relevant to other regions as well. A similar exhaustion of ideological fervor, reversion to pragmatism, and financial pressures from debt are in evidence in many countries and regions. In many respects the debate over nonoffensive defense has potential relevance to areas of conflict outside Europe.

Yet there is one difference between the parties to the European confrontation and those in other areas. In Europe the two military superpowers are themselves in the front lines as direct antagonists. In other regions of conflict, one or both superpowers are hovering at the horizon, with a power to intervene that states may need to factor into their defense calculus beyond their concerns about hostile immediate neighbors. Although the costs to the superpowers of attempting intervention in Third World settings have been rising, it remains an option they are capable of exercising.

EUROPE

So often seen as the paradigm for nonoffensive defense, the NATO/WTO divide presents special problems as well. Even as the political trust and convergence that could facilitate a reciprocal NOD agreement grow, even as verification seems to become attainable, suspicions linger over possible secret development of new military technology that could permit one side to acquire a sudden-strike capability unbeknowst to the other. Weapons modernization presents thorny issues complicating any NOD pact.

Another problem is nuclear weapons. These are variously seen as strictly offensive (governments would presumably shrink from using them on their own territory), as deterring exclusively by prospect of punishment rather than of denial, as political rather than military weapons—but in any event as inconsistent with strict nonoffensive defense doctrine. Yet the military strategies, or at least pretensions, of key members of the Western alliance rely heavily on retention of nuclear forces. Their defense establishments point with satisfaction to the "proven" track record of four decades with nuclear weapons but without war, and express skepticism about whether NOD can substitute so successfully.

Even if the nuclear questions can be finessed, there remain still more difficulties, such as the capability for swift mobilization of reserve forces. Some see the transition to a nonoffensive defense regime as a process proceeding step by step, as with confidence building; but others insist that such step-by-step implementation requires a clear understanding from the outset of what the NOD regime will ultimately entail.

With the Vienna talks on conventional forces in Europe apparently proceeding apace toward an agreement, it appears possible that the desire of Western publics for a lightened defense burden can be satisfied without defense establishments having to make a full-scale restructuring of forces along NOD doctrinal lines—though clearly NOD-ish ideas have made their way even into NATO proposals, like a ban on specific weapons deemed inherently offensive. This raises the question, so fascinating to defense gastronomes, of whether nonoffensive defense involves principles that can be adopted "a la carte," or whether it is a "menu fixe" doctrine whose components are inextricably interwoven and unseverable. The more comprehensive the approach that NOD is expected to be, the less likely seems its adoption as the overarching framework for security.

ASIA

The serving up of a NOD menu in Europe paradoxically puts butterflies in many Asian stomachs. The offensive-capable weapons and forces that a superpower would abandon in the West might not be eliminated but redeployed in the East. There many of the political ingredients needed for common movement toward nonprovocative defense structures are absent.

Japan, the region's largest military power a half-century ago and its largest economic power today, would like to see itself as an idyllic prototype of nonoffensive defense, with strong military forces capable of repelling attacks but incapable of projecting power against an adversary's homeland—in short, a pure case of a major power's "deterrence by denial." However, Japan's security has an important guarantor: the United States, whose nuclear umbrella provides deterrence by punishment. Japanese qualms about the possible movement east of Soviet offensive power underscore Japan's dependence on the more offensively oriented U.S. forces and uncertainty about the efficacy of its own unilateral NOD force structure. If Japan hopes to induce the Soviets to dismantle their offensive capabilities in the Far East and create a viable NOD security framework, its lone bargaining chip is the retaliatory capability of the United States.

By contrast, one area where notions of nonoffensive defense most assuredly have not taken hold—but where the geographic and military setting is in many ways similar to Central Europe—is the Korean peninsula. A bipolar conflict, a balance of huge and well-equipped armed forces, even the mountainous terrain would all make Korea an attractive Asian testing ground for nonoffensive defense strategies. Unfortunately, so long as the political enmity and rejection of legitimacy that each Korean state directs toward the other continues, any progress toward reducing tensions and force levels is extremely unlikely, whether by NOD or other means.

In Southeast Asia most countries rely on nonoffensive defense by default. By contrast, in South Asia major states like India and Pakistan have large, offense-capable military establishments. Both have other neighbors of whom they feel a need to be wary—in India's case another nuclear-armed state with an even larger population. Yet India seems reconciled to a nonoffensive defense strategy against China aimed at repelling any future incursion into disputed Himalayan territory. By contrast, both India and Pakistan have a broad mix of offensive and defensive capabilities clearly intended toward each other, and despite nearly two decades without war the persistence makes movement to a nonoffensive defense regime between them unlikely.

MIDDLE EAST AND CENTRAL AMERICA

The two regions where the nonoffensive defense model seems most implausible are ironically the ones where its prototypes are arguably being tested. This discovery—reminiscent of the delighted surprise of Moliere's bourgeois gentilhomme on learning he had all along been composing his thoughts in prose—suggests the elasticity as well as a potential broad applicability of the NOD idea.

It is obvious to all why NOD cannot work in the Arab-Israeli conflict. Nowhere are underlying passions and animosities so deep, the denial of legitimacy so fervent, the strategic depth of territory in which the defender could absorb an attack so slight. Yet the Camp David treaty between Egypt and Israel, with its detailed provision of force levels and configurations within prescribed distances of the border and with its verification system, prefigures the NOD debate in Europe. Even with political relations that are cooler and more prickly than those between East and West, Egypt and Israel have demonstrated the durability of such an agreement. The difficulty, of course, lies in finding enough political agreement with the other parties to the conflict to allow a renunciation of force.

In the case of Central America, the Esquipulas accord now being implemented provides a framework for security among several ideologically divided small states involving troop reductions and deployments that are guaranteed by outside verification teams. The low technology armies of Central American peasant conscripts admittedly do not present the military complexities that bedevil negotiators in Vienna, and it is possible that little in the Esquipulas experience can be recycled back to inform Europe's planning for a new defense order. Still, it may be a useful paradigm for conflict reduction in other Third World settings.

NOD AND THE MULTILATERAL ARENA

The prospect of a conversion of national and alliance defense strategies to nonoffensive models would suggest the possibility of some measurable progress toward disarmament goals inscribed decades ago in the United Nations Charter. But the role for multilateral institutions in advancing NOD strategic concepts is not obvious. Where nations judge their security can be assured by nonoffensive defense strategies on a reciprocal basis, the standards for NOD observance will inevitably reflect the particular circumstances of the specific region in question. There may be cases where mutual NOD-ed assent is tied to outside verification measures, as has been the case with both the Sinai and Central America. In such cases the parties to the agreement are best able to decide where they will repose their confidence—perhaps in a UN mission, perhaps that of a regional organization or of a particular "honest broker" third party. At this early stage in the development of NOD as a usable defense strategy, nothing would more quickly scuttle it than its grandiose universalization.

List of Contributors:

Dr. J. Soedjati Djiwandono
Vice chairman of the supervisory board
Indonesia Centre for Strategic and International Studies, Jakarta

Yuri Fyodorov
Senior researcher, Institute of World Economy and International Relations
(IMEMO), Moscow
USSR

Alex Gliksman
Project Director
Twentieth Century Fund
USA

Lic. Luis Herrera-Lasso
Director
Centro latinoamericano de estudios estratégicos, Mexico D.F.
Mexico

Dr. Andrzej Karkoszka
Research fellow
Polish Institute of International Affairs, Warsaw
Poland

Dr. Jeffrey Laurenti
Executive Director
Multilateral Studies Program of the United Nations Association of the United States
of America (UNA-USA), New York
USA

Dr. Albrecht A. C. von Müller
Executive Director
European Center for International Security, Feldafing
Federal Republic of Germany

Dr. Gwyn Prins
Director
Global Security Programme at the University of Cambridge, Cambridge
United Kingdom

Prof. Tomohisa Sakanaka
School of International Politics, Economics and Business at the Aoyama Gakuin University, Tokyo
Japan

Agha Shahi
President
Islamabad Council of World Affairs, Islamabad
Pakistan

Dr. Jasjit Singh
Director
Institute for Defense Studies and Analyses, New Dehli
India

Dr. Häkan Wiberg
Director
Centre of Peace and Conflict Research at the University of Copenhagen, Copenhagen
Denmark

For Product Safety Concerns and Information please contact our EU
representative GPSR@taylorandfrancis.com
Taylor & Francis Verlag GmbH, Kaufingerstraße 24, 80331 München, Germany

www.ingramcontent.com/pod-product-compliance
Lightning Source LLC
Chambersburg PA
CBHW050709280326
41926CB00088B/2904

9 780367 628260